# SpringerBriefs in Electrical and Computer Engineering

For further volumes:
http://www.springer.com/series/10059

Leandro Balby Marinho • Andreas Hotho
Robert Jäschke • Alexandros Nanopoulos
Steffen Rendle • Lars Schmidt-Thieme
Gerd Stumme • Panagiotis Symeonidis

# Recommender Systems for Social Tagging Systems

 Springer

Leandro Balby Marinho
Federal University of Campina Grande
Brazil
lbmarinho@dsc.ufcg.edu.br

Robert Jäschke
University of Kassel
Germany

Steffen Rendle
University of Konstanz
Germany

Gerd Stumme
University of Kassel
Germany

Andreas Hotho
University of Würzburg
Germany

Alexandros Nanopoulos
University of Hildesheim
Germany

Lars Schmidt-Thieme
University of Hildesheim
Germany

Panagiotis Symeonidis
Aristotle University
Greece

ISSN 2191-8112          e-ISSN 2191-8120
ISBN 978-1-4614-1893-1   e-ISBN 978-1-4614-1894-8
DOI 10.1007/978-1-4614-1894-8
Springer New York Dordrecht Heidelberg London

Library of Congress Control Number: 2012931154

Printed on acid-free paper

Springer is part of Springer Science+Business Media (www.springer.com)

# Preface

Social tagging systems are Web 2.0 applications that promote user participation through facilitated content sharing and annotation of that content with freely chosen keywords, called *tags*. Despite the potential of social tagging to improve organization and sharing of content, without efficient tools for content filtering and search, users are prone to suffer from information overload as more and more users, content, and tags become available on-line. Recommender systems are among the best known techniques for helping users to filter out and discover relevant information in large datasets. However, social tagging systems put forward new challenges for recommender systems since – differently from the standard recommender setting where users are mainly interested in content – in social tagging systems users may additionally be interested in finding tags and even other users.

The goal of this book is to bring together important research in a new family of recommender systems aimed at serving social tagging systems. While by no means exhaustive, the chapters introduce a wide variety of recent approaches, from the most basic to the state-of-the-art, for providing recommendations in social tagging systems. The focus is on tag recommendations and tag-aware recommendations, which are the prevalent recommendation tasks in the literature and real-world social tagging systems. The material covered in the book is aimed at graduate students, teachers, researchers, and practitioners in the areas of web mining, e-commerce, information retrieval, and machine learning.

The idea for this book emerged from a long history of fruitful cooperation between the authors, who have been actively contributing in many of the topics covered in this book. Many parts of the book are built on top of the authors' previous book chapter entitled *Social Tagging Recommender Systems* published in the *Recommender Systems Handbook* in 2011; which triggered the cooperation with Springer for extending it into a book.

The book is organized into three parts. Part I provides introductory material on social tagging systems and recommender systems. Part II presents a wide variety of recommendation techniques, ranging from the most basic

methods to the state-of-the-art, as well as strategies for evaluating these recommender systems. Part III provides a detailed case study on the technical
aspects of deploying and evaluating recommender systems in BibSonomy, a
real-world social tagging system of bookmarks and scientific references.

# Contents

## Part II Recommendation Techniques for Social Tagging Systems

## Part III Implementing Recommender Systems for Social Tagging

# Part I
# Foundations

# Chapter 1
# Social Tagging Systems

Social Tagging Systems (STS for short) are web applications where users can upload, tag, and share resources (e. g., websites, videos,photos, etc.) with other users. STS promote decentralization of content control and lead the web to be a more open and democratic environment. As we will see in the course of this book, STS put forward new challenges and opportunities for recommender systems, but before we delve into how to design and deploy efficient recommender systems for STS, in this chapter we formally define social tagging systems and their data structures, elaborate on the different recommendation tasks demanded by STS users, introduce real-world STS that already feature recommendation services, and fix the notation we will use throughout the book. The chapter is based on work published in [9].

## 1.1 Introduction

The idea of tagging objects with categories in order to make them more recognizable and understandable was first systematized by Aristotle in his Categories treatise[1], where he analyzes the differences between classes and objects. Since then, categorization has been used for a wide range of different purposes, such as library classification, product catalogs, biological taxonomies, yellow pages of telephone directories, web catalogs, semantic web ontologies, etc. A property shared by most of these classification systems is that there is a restricted and selective number of persons involved in the conception, assignment, and maintenance of the categories. Those persons are usually experts on the respective domain, e. g., biologists for biological taxonomies and librarians for document categorization. However, with the ad-

---

[1] An English translation of the original Aristotle treatise is provided by E. M. Edghill at http://www.classicallibrary.org/aristotle/categories/index.htm.

vent of social tagging systems, the democratization of content creation and categorization enabled ordinary users to become the "experts" themselves.

Social tagging systems are Web 2.0 applications concerned with the publication and tagging of web resources by ordinary internet users. These systems are now widespread, with millions of people using them daily to organize and retrieve on-line content. STS, such as Delicious,[2] BibSonomy,[3] Flickr,[4] Last.fm,[5] etc., bring people together through their shared interests, e. g., music in Last.fm, photos in Flickr, and scientific publication references and bookmarks in BibSonomy. In STS users can upload resources, e. g., URLs of websites in Delicious, BIBTEX entries in BibSonomy, photos in Flickr, sound tracks in Last.fm, etc., and annotate them with a list of freely chosen keywords typically called *tags*. Although the primary goal of tags is to help individual users to organize and retrieve their own content, the exposition of tags by the system ends up benefiting other users since they can adopt each other's tags for browsing and annotating resources. With the increase of tagging activity, a lightweight collaborative classification system, typically known as *folksonomy*,[6] emerges. STS have raised a lot of attention recently due to their potential to improve search and personal organization of resources, while introducing new opportunities for data mining and new forms of social interaction.

This chapter is structured as follows: In Section 1.2 we present a formal model of folksonomies and in Section 1.3 we show how users can navigate the folksonomy through *tag clouds*. In Section 1.4 we present the different data structures of STS. In Section 1.5 we present the recommendation tasks in STS and in Section 1.6 we briefly present some real-world STS that already feature some kind of recommendation service. In Section 1.7 we fix some general notation to be used throughout the book. We finally close with further reading and references.

## 1.2 Folksonomies

Folksonomies are the underlying structure of social tagging systems. They result from the practice of collaboratively creating and managing tags to annotate and categorize content. Tags, in general, are a way of grouping content by category to make it easy to view by topic. This is a grass-root approach to organize a site and help users to find content they are interested in. Formally, a folksonomy is defined as a relational structure $\mathbb{F} := (U, R, T, Y)$ in which

---

[2] http://www.delicious.com/

[3] http://www.bibsonomy.com/

[4] http://www.flickr.com/

[5] http://last.fm/

[6] The term *folksonomy* refers to a blend of the two words *folk* and *taxonomy*.

- $U$, $R$, and $T$ are disjoint non-empty finite sets, whose elements are called users, resources, and tags, respectively, and
- $Y$ is the set of observed ternary relations between them, i. e., $Y \subseteq U \times R \times T$, whose elements are called tag assignments [5].
- A post corresponds to the set of tag assignments of a user for a given resource, i. e., a triple $(u, r, T_{u,r})$ with $u \in U$, $r \in R$, and a non-empty set $T_{u,r} := \{t \in T \mid (u, r, t) \in Y\}$.

Users are typically described by usernames, whereas tags may be arbitrary strings. What is considered a resource depends on the type of the system. For instance, in Delicious, the resources are URLs, in Flickr pictures, in Bib-Sonomy URLs or publication references, and in Last.fm, the resources can be artists, song tracks or albums.

## 1.3 Tag Clouds

*Tag clouds* provide an easy way to navigate the tags, resources, and users of a folksonomy. A tag cloud is a visual representation of user-generated tags, where the popularity of a tag is denoted by its font size, i. e., the larger the font, the more popular the tag. Thus, both finding a tag by alphabet and by popularity is possible, and consequently, the collection of resources that are associated with a tag. Tag clouds first appeared in Flickr [1], being rapidly adopted by other STS, namely Delicious and Technorati,[7] and today are an inherent component of any social tagging system. The idea has been extended to other systems, where the same principle is used for visualizing data, text, and even results of search engines [4]. Figure 1.1 depicts a tag cloud displaying the most popular tags used in Last.fm.

As a side effect, tag clouds may introduce a bias in favor of the most popular tags in the system. It is usually possible to see the tag clouds in all different levels of granularity of the folksonomy, i. e., the tag cloud per user or per resource.

## 1.4 Data Representation

Folksonomy data can be represented in different ways, and as we will see in Part II of this book, different representations stimulate different types of models.

---

[7] http://technorati.com/

**Fig. 1.1** Tag cloud showing the most popular tags used in Last.fm.

### 1.4.1 Folksonomies as Tensors

The set of tag assignments in $Y$ can be represented as third-order *tensor* (3-dimensional array) $\mathcal{A} = (a_{u,r,t}) \in \mathbb{R}^{|U| \times |R| \times |T|}$. There are different ways to represent $Y$ as a tensor. Panagiotis et al. [11], for example, proposed to interpret $Y$ as a binary tensor where 1 indicates observed tag assignments and 0 missing values (see the left-hand side of Figure 1.2):

$$a_{u,r,t} := \begin{cases} 1, & (u,r,t) \in Y \\ 0, & \text{else} \end{cases}$$

Rendle et al. [10], on the other hand, distinguish between positive and negative tag assignments and missing values in order to learn a personalized ranking of tags (see Chapter 4). The idea is that positive and negative examples are only generated from observed tag assignments. Observed tag assignments are interpreted as positive feedback, whereas the non-observed tag assignments of an already tagged resource are negative evidences. All other entries, i.e., all tags for a resource that a user has not tagged yet, are assumed to be missing values (see the right-hand side of Figure 1.2).

### 1.4.2 Folksonomies as Hypergraphs

An equivalent, but maybe more intuitive representation of a folksonomy, is an undirected tripartite hypergraph $G_{\mathbb{F}} := (V, E)$, where $V := U \,\dot\cup\, R \,\dot\cup\, T$ is

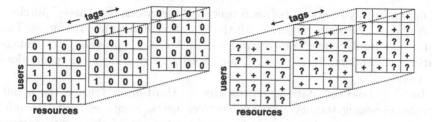

**Fig. 1.2** *Left* [11]: sparse tensor representation where positive feedback is interpreted as 1 and the rest as 0. *Right* [10]: observed tag assignments are considered positive feedback while the non-observed tag assignments for a given already tagged resource are marked as negative feedback. All other entries are missing values [9].

the set of nodes, and $E := \{\{u, r, t\} \mid (u, r, t) \in Y\}$ is the set of hyperedges (see Figure 1.3).

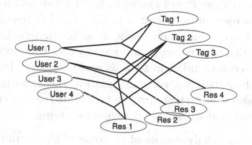

**Fig. 1.3** Tripartite undirected hypergraph representation of a folksonomy [9].

## 1.5 Recommendation Tasks in STS

If on the one hand STS bring new opportunities for improving search and personal organization of resources, they revive old problems on the other, namely the problem of information overload. Millions of individual users and independent providers are flooding STS with content and tags in an uncontrolled way, thereby reducing the user's ability to retrieve relevant information. One of the most successful approaches for increasing the level of relevant content over the "noise" that continuously grows as more and more content becomes available on-line lies on Recommender Systems (RS for short). Recommender systems are applications that help users finding useful objects by automatically reducing the space of choices to the most relevant ones. In the typical scenario, RS algorithms operate over second-order tensors, or matrices, representing binary relations between users and resources, e. g., users' purchase history on an e-commerce site, with the aim of recommending resources. Two notable examples are the e-commerce sites Amazon,[8] providing recommendations of products (e. g., books), and Netflix[9], featuring recommendations of movies.

In STS, however, data is represented as third-order tensors (or hypergraphs) denoting ternary relations between users, resources, and tags, and therefore, recommendations can be provided for any of these entity types. We will refer to RS that can recommend more than one entity type, or mode, as *multi-mode recommender systems* (cf. Chapter 2). Since there are three modes in STS, there are 18 possible recommendation tasks, i. e.,

- $3 \times 3$ tasks where given an entity of class $i$, we want to predict an entity of class $j$ with $i, j \in \{U, R, T\}$(see Figure 1.4). For example, given a target user, recommend other resources he/she hast not yet tagged. We will refer to this kind of recommendation task as *single input*.
- $3 \times 3$ tasks where given an entity of class $i$ and one of class $j$, we want to predict an entity of class $k$. For example, given a user and a resource predict a tag, or given a user and a resource predict another resource. Notice that in this kind of recommendation tasks, most of the existing methods only deal with scenarios where $i \neq j$, and so do we in this book. We will refer to this kind of recommendation task as *multi-input*. In fact, we will focus on two types of multi-input recommendation tasks, namely *tag* and *tag-aware* recommendation, which are the prevalent recommendation tasks in the literature and real-world social tagging systems.

In the following we briefly comment on some of the main recommendation tasks currently supported by real-world social tagging systems.

## *1.5.1 User Recommendation*

Considering single-input, this task predicts other users for a given target user. For example, Last.fm users could be interested in other users sharing similar musical preferences. For multi-input, this task predicts other users given a pair of entities of the other types. For example, Last.fm users could be interested in other users who have also tagged their resources with the tags *rock* and *blues*, for example. By browsing the profiles of similar users, users can improve their ability in discovering relevant content.

---

[8] http://www.amazon.com/
[9] http://netflix.com/

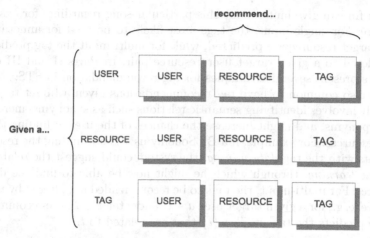

**Fig. 1.4** Single input recommendation tasks on STS. Figure is adapted from Bogers (2009) [2]

## 1.5.2 Resource Recommendation

For single input, the resources most likely to be relevant to a given target user are predicted. For example, a Last.fm user might be interested in receiving recommendations about song tracks that he or she has not yet heard before. Or in the case of multi-input, this same user might want to receive only musical resources tagged with *rock* as recommendations. In Chapters 2 and 3 we provide more details on this recommendation task.

Typically, only the resources that the user has not yet accessed are recommended. The reason is that in most STS users do not pick the same resource twice, e. g., users do not upload the same photo twice in Flickr or do not upload the same scientific reference twice in CiteULike or BibSonomy. This is not the case for tag recommendation, where the users can use the same tag more than once to annotate different resources.

## 1.5.3 Tag Recommendation

Tag recommendation can relieve users from the eventually time consuming task of coming up with a good set of tags, since recognizing which tags to use for annotating a given resource requires far less cognitive effort than conceiving. Another nice property of tag recommendation is that the user can learn about a resource just by looking at the recommendation list. If a user adds a random song, about which he/she has no previous information, to a playlist in Last.fm, for instance, the tag recommendation list provided by Last.fm can give hints about this particular song regarding, for example, its genre. For single input, the tags most likely to be used for annotating a given target resource are predicted, while for multi-input the tag predictions are tailored to a given target user/resource pair. In Parts II and III of the book we present several techniques for tag recommendation in STS.

It is also common to have tag recommendations given a target tag. This typically involves identifying semantic relations such as synonyms, meronyms and hyponyms, and might increase the chances of the user in finding the desired resources. For example, if a BibSonomy user is not finding the resources he wants with the tag *data mining*, the system could suggest the related tag *machine learning*, through which he might now be able to find the desired resources. For multi-input, the tags to be recommended are filtered by user or resource, e. g., for a given target user $u \in U$ and a tag $t \in T$, the recommender system predicts the tags in $T_u \setminus \{t\}$ that are related to $t$.

## 1.6 Recommendations in Social Tagging Systems

Most social tagging systems provide recommendations of some of the afore-mentioned types. However, most often the details of the methods are not publicly available. In this section we give an overview of recommender systems in social tagging systems. Relatively well described are the recommendations provided by BibSonomy and CiteULike. Therefore, we describe them in more detail.

### 1.6.1 BibSonomy

BibSonomy contains a tag recommender since 2006. The first implementation used tags that were extracted from the title combined with the most popular tags of the user [7]. In 2009 a new tag recommendation framework was introduced as cornerstone of the ECML PKDD Discovery Challenge's online tag recommendation task. The framework is presented in Chapter 6, an evaluation of different recommendation methods follows in Chapter 7. After the challenge, the winning recommender [8] was chosen as new tag recommender for BibSonomy. Two simple recommenders function as fallback (cf. Section 6.5.3).

Users get also similar user suggestions. These are computed using different similarity measures (Jaccard, Cosine, TF-IDF, FolkRank). By clicking on such a user, one gets a personalized ranking of the users' posts, based on the overlap of the tag clouds.[10] If one finds the user interesting, one can follow him or her (see Figure 1.5).

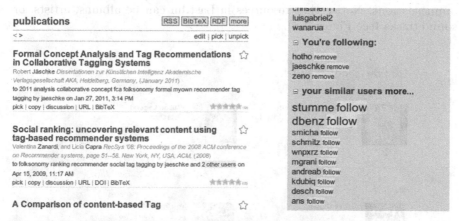

**Fig. 1.5** On the left hand side, the system shows the most recent publications of the followed users, while on the right hand side, it is presented the followed users and the recommended users that might be interesting to follow.

Once a user follows another user, he/she can get a personalized ranking of the other user's posts.

## 1.6.2 CiteULike

CiteULike already supports four of the nine single-input recommendation tasks depicted in Figure 1.4.[11] The most recent addition to CiteULike has been the recommendation of articles in October 2009. It is based on work by Toine Bogers [3] and exploits the historical preferences of users for certain articles and research areas, to locate and recommend relevant articles that are new to the user. Users can copy a recommended article to his/her profile or 'reject' the recommendation. In the first six weeks after the introduction, 9930 articles were rejected and 2323 accepted.[12]

CiteULike also presents a user his/her neighbors based on the overlap between his/her and other user's bookmarked articles. For that list CiteULike filters out users that have less than the median number of articles in common.

---

[10] http://blog.bibsonomy.org/2009/05/new-features-released-similar-users.html

[11] http://blog.citeulike.org/?p=11

[12] http://blog.citeulike.org/?p=136

## *1.6.3 Other Systems*

Last.fm, a social on-line internet radio, features user, resource, and tag recommendations. Notice that resources in Last.fm can be albums, artists, or sound tracks (see Figure 1.6).

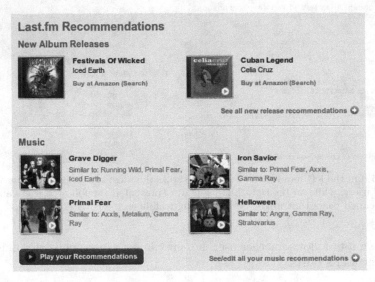

**Fig. 1.6** Resource recommendations in Last.fm.

YouTube, in turn, provides resource and tag recommendations. Delicious recommends web pages and other tags related to a user's tag since 2005.[13] Furthermore, related web pages are recommended for bookmarks. Table 1.1 summarizes the recommendation tasks supported by some of the most popular social tagging systems available.

**Table 1.1** Recommendation modes supported by real-world social tagging systems. A "X" denotes that the system supports recommendations of the corresponding entity class.

| STS | User | Resource | Tag |
|-----|------|----------|-----|
| BibSonomy | X | X | X |
| CiteULike | X | X | X |
| Last.fm | X | X | X |
| Delicious | – | X | X |
| YouTube | – | X | X |

---

[13] http://blog.delicious.com/blog/2005/08/people_who_like.html

## 1.7 Notation

In this section we fix some general notation for the book. For a set $X$ and a (injective) function $f : X \to \mathbb{R}$,

$$\underset{x}{\mathrm{argmax}}^{k} f(x) := X' \subseteq X : |X'| = k, \forall x' \in X', x \in X \setminus X' : f(x') \geq f(x)$$

denotes the set of $k$ largest elements from $X$ (with respect to $f$). If $f$ is not injective, i.e., it could happen that two elements have the same $f$-value, and this definition is not unique, by abuse of notation $\mathrm{argmax}_x^k f(x)$ denotes a random set of $k$ largest elements (i.e., ties are broken at random).
For a statement $A$,

$$\delta(A) := \begin{cases} 1, & \text{if } A, \\ 0, & \text{otherwise} \end{cases}$$

We denote tensors by calligraphic uppercase letters (e.g., $\mathcal{A}$, $\mathcal{B}$), matrices by uppercase letters (e.g., $A$, $B$), scalars by lowercase letters (e.g., $a$, $b$), and vectors by bold lowercase letters (e.g., $\mathbf{a}$, $\mathbf{b}$).
For a vector $\mathbf{x} \in \mathbb{R}^n$,

$$\bar{x} := \frac{1}{n} \sum_{i=1}^{n} x_i$$

denotes the mean value.
For a set $A$,

$$\mathcal{P}(A) := \{B \subseteq A\}$$

denotes the power set of A.

For later convenience, we define several subsets of elements of a folksonomy in Table 1.7.

**Table 1.2** Subset notation for denoting elements of a folksonomy.

| Notation | Description |
| --- | --- |
| $Y_t$ | tag assignments where tag $t \in T$ appears. |
| $Y_{u,t}$ | tag assignments where both user $u \in U$ and tag $t \in T$ appear. |
| $Y_{u,r}$ | tag assignments where both user $u \in U$ and resource $r \in R$ appear. |
| $Y_{r,t}$ | tag assignments where both resource $r \in U$ and tag $t \in T$ appear. |
| $T_u$ | set of tags assigned by a given user $u \in U$. |
| $T_r$ | set of tags assigned to a given resource $r \in R$. |
| $T_{u,r}$ | set of tags assigned to a given resource $r \in R$ by user $u \in U$. |
| $U_r$ | set of users who tagged the resource $r \in R$. |
| $R_u$ | set of resources tagged by user $u \in U$. |
| $R_t$ | set of resources tagged with tag $t \in T$. |
| $R_{u,t}$ | set set of resources tagged by user $u \in U$ with tag $t \in T$. |

The set of recommendations of an entity type $E \in \{U, R, T\}$ of the folksonomy, given a target entity $i$ or a pair of target entities $(i, j)$, is denoted by $\hat{E}_i$ or $\hat{E}_{i,j}$ where $x$ and $y$ are elements of distinct entity types. For example, the set of top-$n$ recommended tags for a target user $u \in U$ and a given resource $r \in R$ is denoted by $\hat{T}_{u,r}$, while the set of recommended resources for a target user $u \in U$ is denoted by $\hat{R}_u$.

## 1.8 Further Reading

The concept of shared on-line resources started with the launch of itList dating back to April 1996 [12]. The service was free and allowed users to store, organize, and share their bookmarks on-line with other users. Within the next few years, other similar services appeared and became competitive, such as Backflip[14] and Blink[15] [6]. Founded in 2003, Delicious was responsible for coining the term social bookmarking and pioneered tagging. Inspired by the increasing popularity of Delicious, several other STS started to appear, supporting different kinds of resources, such as CiteUlike,[16] Connotea,[17] (also called social citation services), and Connectbeam,[18] which included a social tagging service aimed at businesses and enterprises [12].

The issue of multi-mode search in STS was first investigated by Hotho et al. [5], where they formalized the notion of folksonomies and introduced an adaptation of PageRank, called FolkRank, for retrieving users, resources, and tags in social tagging systems. This algorithm was later used for personalized tag recommendations (cf. Chapter 4).

## References

1. Paul Bausch and Jim Bumgardner. *Flickr Hacks*. O'Reilly Press, 2006.
2. Toine Bogers. Science papers that interest you. http://blog.citeulike. org/?p=11. accessed on 07/14/2011.
3. Toine Bogers. *Recommender Systems for Social Bookmarking*. PhD thesis, Tilburg University, Tilburg, The Netherlands, December 2009.
4. Y. Hassan-Montero and V. Herrero-Solana. Improving tag-clouds as visual information retrieval interfaces. In *InScit '06: Proceedings of the*

---

[14] http://www.backflip.com/
[15] http://www.blinkpro.com/
[16] http://www.citeulike.org/
[17] http://www.connotea.org/
[18] http://www.connectbeam.com/

*International Conference on Multidisciplinary Information Sciences and Technologies*, 2006.

5. Andreas Hotho, Robert Jäschke, Christoph Schmitz, and Gerd Stumme. Information retrieval in folksonomies: Search and ranking. In York Sure and John Domingue, editors, *The Semantic Web: Research and Applications*, volume 4011 of *Lecture Notes in Computer Science*, pages 411–426, Berlin/Heidelberg, June 2006. Springer.

6. LaJean Humphries. Extras - itlist and other bookmark managers. Weblog Entry at http://www.llrx.com/extras/itlist.htm, 2000. accessed on 07/14/2011.

7. Jens Illig. Entwurf und Integration eines Item-Based Collaborative Filtering Tag Recommender Systems in das BibSonomy-Projekt. Project report, Fachgebiet Wissensverarbeitung, Universität Kassel, 2006.

8. Marek Lipczak, Yeming Hu, Yael Kollet, and Evangelos Milios. Tag sources for recommendation in collaborative tagging systems. In Folke Eisterlehner, Andreas Hotho, and Robert Jäschke, editors, *ECML PKDD Discovery Challenge 2009 (DC09)*, volume 497 of *CEUR-WS.org*, pages 157–172, 2009.

9. Leandro Balby Marinho, Alexandros Nanopoulos, Lars Schmidt-Thieme, Robert Jäschke, Andreas Hotho, Gerd Stumme, and Panagiotis Symeonidis. *Social Tagging Recommender Systems*, pages 615–644. Springer US, 2011.

10. Steffen Rendle, Leandro B. Marinho, Alexandros Nanopoulos, and Lars Schimdt-Thieme. Learning optimal ranking with tensor factorization for tag recommendation. In *KDD '09: Proceedings of the 15th ACM SIGKDD International Conference on Knowledge Discovery and Data Mining*, pages 727–736. ACM, 2009.

11. Panagiotis Symeonidis, Alexandros Nanopoulos, and Yannis Manolopoulos. Tag recommendations based on tensor dimensionality reduction. In *RecSys '08: Proceedings of the 2008 ACM conference on Recommender systems*, pages 43–50. ACM, 2008.

12. Wikipedia. Social bookmarking. http://en.wikipedia.org/wiki/Social_bookmarking. accessed on 07/14/2011.

# Chapter 2
# Recommender Systems

In the following we will describe systematically and formally the most important problems related to recommender systems and give some references to actual solutions. Our focus here is to describe the general recommender systems setting as a base for social recommender systems. See [11, 3] for a more general introduction to recommender systems and a more thorough overview of the state-of-the-art, respectively.

## 2.1 Rating and Item Prediction

The two most basic recommendation problems are rating prediction and item prediction.

In rating prediction, there are users that rate items (e. g., movies, books, electronic devices, articles, resources in the terminology of social systems etc.) explicitly on some scale, say with the numbers 1 to 5, where 1 denotes the least preferred item and 5 the most preferred one. Given such ratings we would like to predict ratings of users for items they did not rate yet. In the most basic scenario, users and items are treated as entities about which nothing else is known, i. e., as IDs or nominal levels. Formally, there are given

- a set $U$ of users,
- a set $I$ of items,
- a set $\mathcal{R} \subseteq \mathbb{R}$ of ratings, e. g., $\mathcal{R} := \{1, 2, 3, 4, 5\}$,
- a set $\mathcal{D}^{\text{train}} \subseteq U \times I \times \mathcal{R}$ of (user, item, rating) triples,
- a (rating) loss function $\ell : \mathcal{R} \times \mathbb{R} \to \mathbb{R}$ where $\ell(r, \hat{r})$ quantifies how bad it is to predict rating $\hat{r}$ if the actual rating is $r$. A typical choice for the loss is absolute error or squared error:

$$\ell_{AE}(r, \hat{r}) := |r - \hat{r}|, \quad \ell_{SE}(r, \hat{r}) := (r - \hat{r})^2$$

Sought is the prediction of the rating for a user and item, i. e.,

$$\hat{r} : U \times I \to \mathbb{R}$$

s. t. for some test set $\mathcal{D}^{\text{test}} \subseteq U \times I \times \mathcal{R}$ of (user, item, rating) triples (from the same unknown distribution as the train set and not available for the construction of $\hat{r}$) the test risk

$$\text{risk}(\hat{r}; \mathcal{D}^{\text{test}}) := \frac{1}{|\mathcal{D}^{\text{test}}|} \sum_{(u,i,r) \in \mathcal{D}^{\text{test}}} \ell(r, \hat{r}(u,i))$$

is minimal.

In the second problem scenario, item prediction, there are no ratings, but just co-occurrences of users and items, e. g., users may view or buy some of the items. Formally, there are given

- a set $U$ of users,
- a set $I$ of items,
- a set $\mathcal{D}^{\text{train}} \subseteq U \times I$ of (user, item) co-occurrences,
- a (ranking) loss function $\ell : \mathcal{P}(I) \times \mathbb{R}^I \to \mathbb{R}$, e. g., recall at $k$

$$\text{recall}_k(J, \hat{r}) := \frac{1}{|J|} |J \cap \underset{i' \in I}{\text{argmax}}^k \, \hat{r}(i')|, \quad J \subseteq I$$

Sought is for every user a ranking of the items, i. e., a score function

$$\hat{r} : U \to \mathbb{R}^I \quad \text{or equivalently } \hat{r} : U \times I \to \mathbb{R}$$

s. t. for some test set $\mathcal{D}^{\text{test}} \subseteq U \times I$ of (user, item) co-occurrences (from the same unknown distribution as the train set and not available for the construction of $\hat{r}$) the test risk

$$\text{risk}(\hat{r}; \mathcal{D}^{\text{test}}) := \frac{1}{|U(\mathcal{D}^{\text{test}})|} \sum_{u \in U(\mathcal{D}^{\text{test}})} \ell(I_u(\mathcal{D}^{\text{test}}), \hat{r}(u)),$$

with $U(\mathcal{D}) := \{u \in U \mid \exists i \in I : (u,i) \in \mathcal{D}\}, \quad I_u(\mathcal{D}) := \{i \in I \mid (u,i) \in \mathcal{D}\}$

is minimal.

If the score function is injective (or made injective by breaking ties at random), it defines for each user $u$ a linear order over the items by

$$i \prec_u j \quad :\Leftrightarrow \quad \hat{r}(u,i) > \hat{r}(u,j), \quad i,j \in I$$

## 2.2 Rating Prediction as Regression Problem

Recommendation problems such as rating and item prediction can be viewed as instances of broader problem classes. The rating prediction problem basically is an (ordinal) regression problem where an (ordinal/numeric) target variable (rating) should be predicted based on two nominal variables (user, item). Among the specific characteristics of the rating prediction is (i) the high number of levels of each of the two nominal variables (many users, many items) and consequently (ii) the extreme sparsity, i. e., that ratings are observed for only very few user/item pairs. Regression problems of this type are also described lucidly as matrix completion problems where rows and columns of the matrix are indexed by the nominal levels of the two variables and cell values are the ratings, most cells being not observed (see e. g., [6]).

Treating the rating prediction problem in a naive way, say, with binary indicator variables for the nominal levels of users and items and a linear model on these variables, leads to a very simple model

$$\hat{r}(u,i) := \mu + \mu^U(u) + \mu^I(i), \quad \mu \in \mathbb{R}, \mu^U : U \to \mathbb{R}, \mu^I : I \to \mathbb{R}$$

where $\mu$ denotes a global average rating and $\mu^U$ and $\mu^I$ model independent user and item effects (often called user and item bias). This model is unsuited for personalized predictions as it does not catch any user/item interactions. If one would use it, again in a naive way, for ranking items for a given target user, then this ranking would be the same for all users. On the other hand, adding an explicit interaction effect between user and item indicator to the model would lead to $|U| \times |I|$ parameters, as many as there are observations in the completed rating matrix.

Therefore, historically, researchers were looking for other methods to model the rating prediction regression problem. For its simplicity, especially the nearest neighbor model got a lot of attention [7, 20, 23]. Here, rating prediction is viewed as a separate problem for each item. Then the user indicator variable is the only variable remaining. Between users a similarity measure is defined based on their rating vectors, e. g., the Pearson correlation of their jointly rated items

$$\text{sim}(u,v) := \text{corr}_{\text{pearson}}(r|_{u,I_u \cap I_v}, r|_{v,I_u \cap I_v})$$

with

$$I_u := \{i \in I \mid \exists r : (u,i,r) \in \mathcal{D}^{\text{train}}\}, \quad u \in U$$

$$r|_{u,J} := (r(u,j))_{j \in J} \in \mathbb{R}^{|J|}, \quad J \subseteq I$$

$$\text{corr}_{\text{pearson}}(x,y) := \frac{\sum_{i=1}^{n}(x_i - \bar{x})(y_i - \bar{y})}{\sqrt{\sum_{i=1}^{n}(x_i - \bar{x})^2}\sqrt{\sum_{i=1}^{n}(y_i - \bar{y})^2}}, \quad x,y \in \mathbb{R}^n$$

For each target item, then the $k$ nearest neighbors having the target item rated are determined and the rating of a given target user for a given target item is predicted by a $k$-nearest-neighbor rule, e. g.,

$$\hat{r}(u,i) := \mu^U(u) + \sum_{v \in N_{u,i}} \frac{\text{sim}(u,v)}{\sum_{v' \in N_{u,i}} \text{sim}(u,v')}(r(v,i) - \mu^U(v))$$

with

$$N_{u,i} := \underset{v \in U:\ i \in I_v}{\operatorname{argmax}^{k}} \operatorname{sim}(u,v)$$

Alternatively, one could swap the roles of users and items, i. e., decompose the problem into separate problems for each item, define a similarity measure between items based on their rating vector by users and predict ratings by a $k$ nearest neighbor rule on items. Nearest neighbor models often are called collaborative filtering or memory-based models in the context of recommender system problems. If users are the instances, they are called user-based, otherwise item-based.

At the end of the 90s, probabilistic latent class models, especially the aspect model [9, 10] have been developed that allowed a richer modelling of the user/item effects through a set of non-observed classes. These models nowadays can be understood as regularized sparse low-rank matrix factorization models [25]. In sparse low-rank matrix factorization models one associates a latent feature vector $\phi$ with every level of each nominal variable and models the interaction between two such variables by a function of their latent feature vectors, e. g., by their scalar product

$$\hat{r}(u,i) := \mu + \mu^U(u) + \mu^I(i) + \langle \phi^U(u), \phi^I(i) \rangle, \quad \phi^U : U \to \mathbb{R}^k, \phi^I : I \to \mathbb{R}^k$$

These models are called sparse matrix factorization models, because when identifying the rating and latent feature functions with the matrices of their values, the rating matrix can be reconstructed by the product of the feature matrices:

$$\hat{r} := \mu \mathbb{I} + \mu^U \mathbb{1}^T + \mathbb{1}(\mu^I)^T + \phi^U(\phi^I)^T,$$

where $\mathbb{I}$ denotes the $|U| \times |I|$ matrix containing only 1's and $\mathbb{1}$ the vector containing $|U|$ many 1's (or $I$ many 1's, respectively). As most entries of this matrix are not observed, one measures the reconstruction error only on the sparse submatrix of observed entries, i. e.,

$$\ell(r, \hat{r}) := ||W^{\text{train}} \odot (r - \hat{r})||$$

where $W^{\text{train}} \in \mathbb{R}^{|U| \times |I|}$ is a weight matrix, usually

$$W^{\text{train}}_{u,i} := \delta(\exists r : \ (u,i,r) \in \mathcal{D}^{\text{train}}), \quad u \in U, i \in I$$

and $|| \cdot ||$ a matrix norm, e.g.,

$$||A|| := \sum_{i=1}^{n} \sum_{j=1}^{m} A_{i,j}^2, \quad A \in \mathbb{R}^{n \times m}$$

and $\odot$ denotes element-wise matrix multiplication. They are called low-rank because the dimension $k$ of the feature vectors (and thus the rank of the resulting reconstruction $\hat{r}$) is small compared to the dimensions $|U|, |I|$ of the original matrix. The models are called regularized, as not just the training loss is minimized, but a combination of training risk and a regularization term, e. g., Tikhonov regularization

$$\text{min. } f(\phi^U, \phi^I) := \text{risk}(\hat{r}; \mathcal{D}^{\text{train}}) + \lambda(||\phi^U||^2 + ||\phi^I||^2), \quad \lambda \in \mathbb{R}_0^+$$

Sparse low-rank matrix factorization models are the state-of-the-art models at the time of writing [13]. They usually provide better performance than other models, do not require to have all training data available at prediction time and are easy to train. For training, different learning methods have been researched. Extremely simple and among the fastest training methods is stochastic gradient descent [13]. Here, one triple $(u, i, r)$ at a time is sampled from the training data and the features are updated along the negative gradient with some learning rate $\eta \in \mathbb{R}^+$ until convergence:

$$\phi^U(u) := \phi^U(u) - \eta \frac{\partial \ell}{\partial \hat{r}}(r, \hat{r}(u, i)) \frac{\partial \hat{r}}{\partial \phi^U(u)}(u, i) - 2\eta\lambda\phi^U(u)$$

$$\phi^I(i) := \phi^I(i) - \eta \frac{\partial \ell}{\partial \hat{r}}(r, \hat{r}(u, i)) \frac{\partial \hat{r}}{\partial \phi^I(i)}(u, i) - 2\eta\lambda\phi^I(i)$$

So for example, for the squared error loss this simply yields

$$\phi^U(u) := \phi^U(u) - 2\eta(r - \hat{r}(u, i))\phi^I(i) - 2\eta\lambda\phi^U(u)$$

$$\phi^I(i) := \phi^I(i) - 2\eta(r - \hat{r}(u, i))\phi^U(u) - 2\eta\lambda\phi^I(i)$$

Extensions of the simple matrix factorization model also can cope with the ordinal level of the rating variable [15].

To demonstrate the usefulness of personalized models such as nearest neighbor models or matrix factorization models in some specific domain, one usually compares them with non-personalized models. More exactly, non-personalized models are models that are constant, either the globally constant model

$$\hat{r}(u, i) := \mu, \quad \mu \in \mathbb{R}$$

or constant w. r .t. one of the user or item variable, i. e., user or item averages:

$$\hat{r}(u, i) := \mu + \mu^U(u), \quad \mu \in \mathbb{R}, \mu^U : U \to \mathbb{R}$$

$$\hat{r}(u, i) := \mu + \mu^I(i), \quad \mu \in \mathbb{R}, \mu^I : I \to \mathbb{R}$$

## 2.3 Item Prediction as Ranking Problem

The item prediction problem often is viewed as a set-valued classification problem (usually called a multi-label classification problem). As such it could be described as a set of dependant binary classification problems, one for

each item. A naive Perceptron model again using binary indicator variables
for each level of the user and item variable looks like this:

$$\hat{r}(u,i) := \mu + \mu^U(u) + \mu^I(i), \quad \mu \in \mathbb{R}, \mu^U : U \to \mathbb{R}, \mu^I : I \to \mathbb{R}$$

It suffers from the very same defect as the naive rating prediction model: it
is not personalized, i. e., there is no interaction term for users and items and
such a term cannot be inserted into the model for all interactions as these
parameters are exactly the output one is trying to learn.

Also for item prediction, nearest neighbor models have been used very
successfully very early on. Similarities no longer can be described by the
correlation of the joint rating vectors, but, e. g., by measures of the overlap
of the item sets for two users, e. g., the Jaccard coefficient

$$\text{sim}(u,v) := \frac{|I_u \cap I_v|}{|I_u \cup I_v|}$$

The neigborhood of a user now does not depend on him having rated the
target item, but just on similarity, and the nearest-neighbor rule counts the
fraction of neighbors with the target item:

$$\hat{r}(u,i) := \sum_{v \in N_u} \frac{\text{sim}(u,v)}{\sum_{v' \in N_u} \text{sim}(u,v')} \delta((v,i) \in \mathcal{D}^{\text{train}})$$

with

$$N_u := \underset{v \in U}{\text{argmax}}^{k}\, \text{sim}(u,v)$$

To better understand the item prediction problem, in our opinion three
ideas have been crucial: (i) the problem has been tackled by a probabilistic
latent class model, the aspect model, closely related to the one used for
rating prediction [9, 10]. (ii) the decomposition by a binary classification
model per item has been found not to work, as typical recommender data
sets have disjoint train and test item sets for the same user, i. e., no repeating
items, while pairwise decompositions have been shown to work well [22]. (iii)
the matrix factorization approach and the direct optimization of a ranking
loss have been applied to the item prediction problem [26]. Nowadays the
simplest and most elegant formulation of the item prediction problem is not
as a classification problem, but as a ranking problem using pairs of positive
items (in the train set) and negative items (not in the train set) as pairwise
input, optimizing a simple ranking loss such as AUC

$$\ell_{\text{AUC}}(J, \hat{r}) := \frac{1}{|J||I \setminus J|} \sum_{j \in J, i \in I \setminus J} \delta(\hat{r}(j) > \hat{r}(i)), \quad J \subseteq I$$

and using matrix factorization as the ranking function [19].

By approximating the discontinuous step function $\delta$, e. g., by the logistic function

$$\sigma(x) := \frac{1}{1 + e^{-x}}$$

one gets a differentiable (logarithmic) loss

$$\ell(r, \hat{r}) := \tau(r - \hat{r}), \quad \tau(x) := \ln \sigma(x)$$

that can be optimized directly using a stochastic gradient algorithm on triples $(u, j, i)$ of users $u \in U$, positive items $j \in I_u$ and negative items $i \in I \setminus I_u$.

$$\phi^U(u) := \phi^U(u) - \eta \tau'(\hat{r}(j) - \hat{r}(i))(\phi^I(j) - \phi^I(i)) - 2\eta\lambda\phi^U(u),$$
$$\phi^I(j) := \phi^I(j) - \eta \tau'(\hat{r}(j) - \hat{r}(i))\phi^U(u) - 2\eta\lambda\phi^I(j)$$
$$\phi^I(i) := \phi^I(i) + \eta \tau'(\hat{r}(j) - \hat{r}(i))\phi^U(u) - 2\eta\lambda\phi^I(i)$$

with

$$\tau'(x) := 1 - \sigma(x)$$

This model is known as Bayesian Personalized Ranking (BPR) [19].

## 2.4 User and Item Attributes

In practice, the assumption that users and items are entities about which there is nothing else known often is too restrictive and does not make use of some descriptive information about them. For example, in e-commerce, there are a lot of attributes of the items (products) easily available and there are some known attributes of users (customers). In these scenarios we say we have attributes of users or items. We model them by functions

$$a^U : U \to \mathbb{R}^{n_U}, \quad \text{and} \quad a^I : I \to \mathbb{R}^{n_I}$$

respectively.

Early on models have been developed that partition the recommendation problem into (independent) subproblems for each user, trying to predict the rating or item choice based solely on the item attributes (content-based filtering [5]). As this completely disregards all collaborative information from other users, such models provide useful results only in specific settings where no such information is available (see the new item problem in Section 2.5) or as component models in ensembles (sometimes called hybrid models in the context of recommender systems).

Nowadays, user and item attributes are understood as a second auxiliary relation in a multi-relational setting. As the user and item attribute relation share a nominal variable with many levels with the target relation, it makes sense to factorize both relations and share the features of the shared variable.

The resulting models are called multi-relational matrix factorization models [24, 14]. For the case of user and item attributes, the loss of such a model looks like

$$\ell(\phi^U, \phi^I, \phi^{AU}, \phi^{AI}) := \ell(r, \phi^U(\phi^I)^T) + \lambda_{AU}\ell(a^U, \phi^U(\phi^{AU})^T)$$
$$+ \lambda_{AI}\ell(a^I, \phi^I(\phi^{AI})^T), \quad \lambda_{AU}, \lambda_{AI} \in \mathbb{R}_0^+$$

Such a model could be easily learned again by stochastic gradient descent, sequentially sampling tuples from the different input matrices. The effect of factorizing an auxiliary relation could be understood as data-dependent regularization as we push the latent features $\phi^U$ in a direction where they can be used to reconstruct not just the target rating or item choice relation, but also the auxiliary user attribute relation. The weights $\lambda_{AU}$ and $\lambda_{AI}$ determine how strong this regularization effect should be. As any other regularization parameters they have to be learned as hyperparameters.

Auxiliary information such as user and item attributes obviously are only useful when there is not too much primary information, i.e., ratings or item choices. For some large real recommendation scenarios it has been shown experimentally that collaborative models based on 10 ratings about an item provide better predictions than content-based models on thousands of attributes [16]. So user and item attributes mostly are been useful for users and items that recently joined a system and for whom/which only little rating/item choice information is available (recent user / recent item problems) or none at all (new user / new item problems; see next section).

## 2.5 New User and New Item Problems

A specific class of problems in recommender systems are the so-called new user or new item problems (also called cold-start problems). A new user problem describes the situation of a new user entering the system, so that this user did not yet have rated or choosen any items. In this case obviously none of the personalized models discussed so far could provide any recommendations. In practice these problems cover important cases: new users should not be scared away by getting bad or no recommendations in the beginning, and new items should not have to wait until they are found and taken up by users by chance.

For new users, one could resort to content-based filtering, i.e., to build a separate model for each item that predicts the rating / item choice as function of the user attributes. Better models have been researched where such content-based models are mediated by collaborative models for the ratings / item choices [21].

Besides the question how to deal with new users and new items, the active learning scenario is of interest for recommender systems, i.e., ratings about which items to ask a new or recent user so that his preferences could be learned as quickly as possible [17, 8].

## 2.6 Context-aware and Multi-Mode Recommendations

Traditionally, recommender systems describe an interaction between two entities, users and items. In many scenarios further entities may moderate this interaction and influence a users preference for an item, e. g., the mood of the target user, the actual location, the actual time, the task the target user is pursueing, the group the target user is with, etc. These further circumstances or modes initially have been described in the literature by different names (location-aware recommendations, time-aware recommendations, group recommendations, etc.), but now often collectively are called context-aware recommendations, where the context could be the mood, location, time, etc. Abstracting from the names of the different entities, this problem could be described as a multi-mode recommendation problem (in the literature usually called multidimensional [2, 4]). Formally, the multi-mode rating prediction is as follows: given

- a set $\mathcal{E}$ of entity classes, where each entity class $E \in \mathcal{E}$ is a set of entity instances (i. e., a set of users, items, moods, etc.),
- a set $\mathcal{R} \subseteq \mathbb{R}$ of ratings, e. g., $\mathcal{R} := \{1, 2, 3, 4, 5\}$,
- a set $\mathcal{D}^{\text{train}} \subseteq \prod_{E \in \mathcal{E}} E \times \mathcal{R}$ of (entity$_1$, entity$_2$, ..., entity$_{|\mathcal{E}|}$, rating) tuples,
- a (rating) loss function $\ell : \mathcal{R} \times \mathbb{R} \to \mathbb{R}$ where $\ell(r, \hat{r})$ quantifies how bad it is to predict rating $\hat{r}$ if the actual rating is $r$.

Sought is the prediction of the rating for an entity instance of each class, i. e.,

$$\hat{r} : \prod_{E \in \mathcal{E}} E \to \mathbb{R}$$

s. t. for some test set $\mathcal{D}^{\text{test}} \subseteq \prod_{E \in \mathcal{E}} E \times \mathcal{R}$ of (entity$_1$, entity$_2$, ..., entity$_{|\mathcal{E}|}$, rating) tuples (from the same unknown distribution as the train set and not available for the construction of $\hat{r}$) the test risk

$$\text{risk}(\hat{r}; \mathcal{D}^{\text{test}}) := \frac{1}{|\mathcal{D}^{\text{test}}|} \sum_{(e_1, \ldots, e_{|\mathcal{E}|}, r) \in \mathcal{D}^{\text{test}}} \ell(r, \hat{r}(e_1, \ldots, e_{|\mathcal{E}|}))$$

is minimal.

The item prediction problem can be generalized in the same way to a multi-mode item prediction problem. Specifically for item prediction, it sometimes is interesting to predict another mode than the item mode. As we have seen in section 1.2, in social tagging systems, tags can be described as a third mode, and there it will be interesting to predict for a given user and item (resource), which tags he is likely to use for the item. Any multi-modal item prediction model obviously can be used to predict any mode by just subsituting mode names.

If all modes are nominal (as the user and item modes), a multi-mode recommendation problem can be modeled by means of factorization models of higher order, so called tensor factorization models [18, 12].

Related to, but different from the multi-mode rating recommendation problem is the multi-criteria recommendation problem [1]. Here, the rating is not just a single (ordinal) overall rating, but a compound rating reflecting different criteria or different aspects of the item individually, e. g., for movies suspense, emotion and humor, or for holiday accommodations location, comfort, friendliness, etc. The difference between multi-criteria recommendation problems and multi-mode recommendation problems is that in the latter all modes are observed for a test case, but for multi-criteria recommendations ratings for other criteria are not observed, i. e., not available to base the prediction upon.

If a problem is described as a multi-mode or a multi-criteria problem reflects a specific requirement of the application. For example, item recommendation in a social tagging system has useful (but different) applications as both, as multi-mode recommendation problem and as multi-criteria recommendation problem. As multi-mode recommendation problem we try to predict for a given user and a given set of tags, which items the user may be looking for. Here, the tags may describe the context in which the user is looking for items. On the other hand, as multi-criteria recommendation problem we are treating the tags as a (nominal) rating, so for a given user we are looking for interesting items (and eventually in parallel for tags he may later associate with that item).

For reference in the remaining chapters, we instantiate the context-aware multi-mode item recommendation problem for tags in social systems, for short called tag recommendation: given

- sets $U, R$, and $T$ of users, resources, and tags, respectively,
- a set $Y := \mathcal{D}^{\text{train}} \subseteq U \times R \times T$ of user/resource/tag triples,
- a (ranking) loss function $\ell : \mathcal{P}(T) \times \mathbb{R}^T \to \mathbb{R}$, e. g., recall at $k$

$$\text{recall}_k(S, \hat{s}) := \frac{1}{|S|} |S \cap \operatorname*{argmax}_{t' \in T}^{k} \hat{s}(t')|, \quad S \subseteq T$$

Sought is for every user/resource pair a ranking of the tags, i. e., a score function

$$\hat{s} : U \times R \to \mathbb{R}^T \quad \text{or equivalently } \hat{s} : U \times R \times T \to \mathbb{R}$$

s. t. for some test set $\mathcal{D}^{\text{test}} \subseteq U \times R \times T$ of user/resource/tag triples (from the same unknown distribution as the train set and not available for the construction of $\hat{s}$) the test risk

$$\text{risk}(\hat{s}; \mathcal{D}^{\text{test}}) := \frac{1}{|UR(\mathcal{D}^{\text{test}})|} \sum_{(u,r) \in UR(\mathcal{D}^{\text{test}})} \ell(T_{u,r}(\mathcal{D}^{\text{test}}), \hat{s}(u,r)),$$

with $UR(\mathcal{D}) := \{(u,r) \in U \times R \mid \exists t \in T : (u,r,t) \in \mathcal{D}\}$,

$\quad T_{u,r}(\mathcal{D}) := \{t \in T \mid (u,r,t) \in \mathcal{D}\}$

is minimal.

# References

1. G. Adomavicius and Y. O Kwon. New recommendation techniques for multicriteria rating systems. *IEEE Intelligent Systems*, page 48–55, 2007.
2. G. Adomavicius and A. Tuzhilin. Multidimensional recommender systems: a data warehousing approach. *Electronic Commerce*, page 180–192, 2001.
3. G. Adomavicius and A. Tuzhilin. Toward the next generation of recommender systems: A survey of the state-of-the-art and possible extensions. *IEEE transactions on knowledge and data engineering*, page 734–749, 2005.
4. G. Adomavicius and A. Tuzhilin. Context-aware recommender systems. In *Proceedings of the 2008 ACM conference on Recommender systems RecSys 08*, volume 16, 2008.
5. M. Balabanović and Y. Shoham. Fab: content-based, collaborative recommendation. *Communications of the ACM*, 40(3):66–72, 1997.
6. E. J Candès and B. Recht. Exact matrix completion via convex optimization. *Foundations of Computational Mathematics*, 9(6):717–772, 2009.
7. D. Goldberg, D. Nichols, B. M Oki, and D. Terry. Using collaborative filtering to weave an information tapestry. *Communications of the ACM*, 35(12):61–70, 1992.
8. A. S Harpale and Y. Yang. Personalized active learning for collaborative filtering. In *Proceedings of the 31st annual international ACM SIGIR conference on Research and development in information retrieval*, page 91–98, 2008.
9. T. Hofmann and J. Puzicha. Latent class models for collaborative filtering. In *International Joint Conference on Artificial Intelligence*, volume 16, page 688–693, 1999.
10. Thomas Hofmann. Latent semantic models for collaborative filtering. *ACM Transactions on Information Systems*, 22(1):89–115, January 2004.
11. Dietmar Jannach, Markus Zanker, Alexander Felfernig, and Gerhard Friedich. *Recommender Systems: An Introduction*. Cambridge University Press, 1 edition, September 2010.

12. A. Karatzoglou, X. Amatriain, L. Baltrunas, and N. Oliver. Multiverse recommendation: n-dimensional tensor factorization for context-aware collaborative filtering. In *Proceedings of the fourth ACM conference on Recommender systems*, page 79–86, 2010.
13. Yehuda Koren, Robert Bell, and Chris Volinsky. Matrix factorization techniques for recommender systems. *Computer*, 42(8):37, 30, 2009.
14. C. Lippert, S. H Weber, Y. Huang, V. Tresp, M. Schubert, and H. P Kriegel. Relation-prediction in multi-relational domains using matrix-factorization. In *NIPS Workshop: Structured Input-Structured Output*, 2008.
15. Ulrich Paquet, Blaise Thomson, and Ole Winther. A hierarchical model for ordinal matrix factorization. *Statistics and Computing*, June 2011.
16. I. Pilászy and D. Tikk. Recommending new movies: even a few ratings are more valuable than metadata. In *Proceedings of the third ACM conference on Recommender systems*, page 93–100, 2009.
17. A. M Rashid, I. Albert, D. Cosley, S. K Lam, S. M McNee, J. A Konstan, and J. Riedl. Getting to know you: learning new user preferences in recommender systems. In *Proceedings of the 7th international conference on Intelligent user interfaces*, page 127–134, 2002.
18. Steffen Rendle. *Context-Aware Ranking with Factorization Models*. Springer Berlin Heidelberg, 1st edition, November 2010.
19. Steffen Rendle, Christoph Freudenthaler, Zeno Gantner, and Schmidt-Thieme Lars. BPR: Bayesian personalized ranking from implicit feedback. In *Proceedings of the Twenty-Fifth Conference on Uncertainty in Artificial Intelligence*, UAI '09, pages 452–461, Arlington, Virginia, United States, 2009. AUAI Press.
20. P. Resnick, N. Iacovou, M. Suchak, P. Bergstrom, and J. Riedl. GroupLens: an open architecture for collaborative filtering of netnews. In *Proceedings of the 1994 ACM conference on Computer supported cooperative work*, page 175–186, 1994.
21. A. I Schein, A. Popescul, L. H Ungar, and D. M Pennock. Methods and metrics for cold-start recommendations. In *Proceedings of the 25th annual international ACM SIGIR conference on Research and development in information retrieval*, page 253–260, 2002.
22. L. Schmidt-Thieme. Compound classification models for recommender systems. In *Fifth IEEE International Conference on Data Mining (ICDM'05)*, pages 378–385, Houston, TX, USA, 2005.
23. U. Shardanand and P. Maes. Social information filtering: algorithms for automating "word of mouth". In *Proceedings of the SIGCHI conference on Human factors in computing systems*, page 210–217, 1995.
24. Ajit P. Singh and Geoffrey J. Gordon. Relational learning via collective matrix factorization. In *Proceeding of the 14th ACM SIGKDD international conference on Knowledge discovery and data mining*, pages 650–658, Las Vegas, Nevada, USA, 2008. ACM.

25. N. Srebro and T. Jaakkola. Weighted low-rank approximations. In *Proceedings of the 25th international conference on Machine learning*, volume 20, pages 720–727, Helsinki, Finland, 2003.
26. M. Weimer, A. Karatzoglou, Q. Le, and A. Smola. Cofi rank-maximum margin matrix factorization for collaborative ranking. *Advances in neural information processing systems*, 20:1593–1600, 2008.

# Part II
# Recommendation Techniques for Social Tagging Systems

# Chapter 3
# Baseline Techniques

In this chapter we introduce the most basic techniques for recommendations in STS. Despite their simplicity, these methods are very easy to implement, cheap to compute, and have proven to attain reasonably good results; features that make them good alternatives to start with by anyone planning on deploying recommendation services in STS.

## 3.1 Constant Models

The most simple recommender systems are those based on counting frequencies of occurrences or co-occurrences of some given entity (or entities) in the data. Although typically regarded as baselines, these recommenders are good alternatives on their own right since they are very cheap to compute and can work with a minimal amount of data. In the following we present several of the recommendation tasks defined in Section 1.5 by means of counting-based techniques.

### 3.1.1 Tag Recommendation

Recommending the most frequent tags of the folksonomy is the most simplistic approach. We will refer to this method as *constant tags*, as it always recommends the same set of tags regardless the target entities given as starting point. For recommending tags, for any user $u \in U$ and any resource $r \in R$, first a linear scan is done in $Y$ for counting each tag's frequency of occurrence. So in this case the prediction function is given by

$$\hat{s}(u, r, t) := |Y_t|$$

Notice that once we have defined a prediction function, recommending the top-$n$ entities of interest boils down to sorting these entities in decreasing order of their scores and selecting the top-$n$. Therefore, for generating the tag recommendation set $\hat{T}_{u,r}$ for any user/resource pair, the tags are sorted in descending order of their predicted scores, and the top-$n$ tags are selected for recommendation. These steps are summarized in the equation

$$\hat{T}_{u,r} := \underset{t\in T}{\arg\max}\,^{n}\,\hat{s}(u,r,t). \tag{3.1}$$

Alternatively, one can score a tag by counting the frequency of co-occurrence of this tag with a given resource (or user), i. e.,

$$\hat{s}(u,r,t) := |Y_{r,t}| \text{ (or } |Y_{u,t}|). \tag{3.2}$$

### 3.1.2 User/Tag-aware Recommendation

The same idea can be extended trivially for recommending users or resources. For predicting the score of a user $u' \in U$ for a given target user $u \in U$, with $u \neq u'$, we can simply count the number of tags (or resources) that the pair of users $(u, u')$ have in common, i. e., $|T_u \cap T_{u'}|$ (or $|R_u \cap R_{u'}|$).

Resource recommender systems that incorporate tags in the recommendation model are usually referred to as *tag-aware recommender systems* [15]. For computing the score of a resource $r \in R$ given a tag $t \in T$, for example, we can just count how often they co-occur in $Y$, i. e., $|Y_{r,t}|$.

### 3.1.3 Remarks on Complexity

As mentioned in the beginning of this section, the methods based on counting usually require very modest computational costs. For computing *constant tags* recommendations, for example, we just need to do a linear scan in $Y$ in order to count the frequencies of occurrences of tags, and then sort the tags by their counts. This results in the following cost:

$$O(|Y| + |T|\log(n)) \tag{3.3}$$

where $n$ is the number of tags to recommend. Note that user and resource recommendation follow the same principles and thus have similar costs.

## 3.2 Projection Matrices

Because of the ternary relational nature of STS data, many recommendation algorithms originally designed to operate on matrices cannot be applied directly, unless the set $Y$ of ternary relations is broken into sets of
binary relations. Marinho and Schmidt-Thieme [8], for example, considered
2-dimensional projections of the original ternary relational data in order to
apply user-based CF for tag recommendations. The idea is to consider either
the projection matrix of $Y$ restricted to users and resources, i. e.,

$$X := \pi_{UR} Y \in \mathbb{R}^{|U| \times |R|}$$

where $x_{u,r} \in \mathbb{R}$ is a real value denoting the strength of the relation between
user $u$ and resource $r$ and $\pi$ denotes the unary projection operator; or the
projection matrix of $Y$ restricted to users and tags, i. e.,

$$X := \pi_{UT} Y \in \mathbb{R}^{|U| \times |T|}$$

Notice that one could also use the projection $\pi_{RT} Y \in \mathbb{R}^{|R| \times |T|}$ restricted to
resources and tags. However, this projection discards the user information and
leads to non-personalized content-based recommendations. While content-
based methods are important in many scenarios, as mentioned in Chapter 2,
in this book we will focus on personalized recommendations.

Most STS do not support ratings, hence the values that $x_{u,r}$ (or $x_{u,t}$)
assume in practice are: (i) binary values, i. e., 1 if a user co-occurred with
a resource (or a tag) and 0 else; (ii) the frequencies which users accessed
resources or assigned tags to their resources; or weighting functions, such
as TF-IDF [7], applied to users, tags, or resources. Figure 3.1 depicts the
2-dimensional projections derived from $Y$.

In practice, the matrix $X$ is represented as a sparse user-resource (or user-
tag) matrix $X \in \mathbb{R}^{|U| \times |R|} \cup \{.\}$ (or $X \in \mathbb{R}^{|U| \times |T|} \cup \{.\}$), where $\{.\}$ denotes
missing values. The matrix $X$ can be decomposed into row vectors:

$$X := (\mathbf{x}_1, \ldots, \mathbf{x}_{|U|})^{\mathrm{T}} \text{ with } \mathbf{x}_u := (x_{u,1}, \ldots, x_{u,|R|}) \text{ or}$$
$$\mathbf{x}_u := (x_{u,1}, \ldots, x_{u,|T|}) \text{ for } u := 1, \ldots, |U|$$

Each row vector $\mathbf{x}_u$ thus corresponds to a user profile representing the
preferences of a particular user for resources (or tags). This decomposition
usually leads to algorithms that leverage user-user similarities such as the
well known user-based collaborative filtering (cf. Chapter 2). These matrices
can alternatively be represented by their column vectors:

$$X := (\mathbf{x}_1, \ldots, \mathbf{x}_{|R|}) \text{ with } \mathbf{x}_r := (x_{1,r}, \ldots, x_{|U|,r})^{\mathrm{T}}, \text{ for } r := 1, \ldots, |R| \text{ or}$$
$$X := (\mathbf{x}_1, \ldots, \mathbf{x}_{|T|}) \text{ with } \mathbf{x}_t := (x_{1,t}, \ldots, x_{|U|,t})^{\mathrm{T}}, \text{ for } t := 1, \ldots, |T|$$

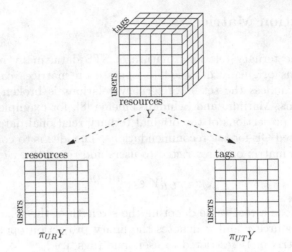

**Fig. 3.1** Projections of $Y$ into the user's resource and user's tag spaces.

This representation leverages resource-resource, or tag-tag, similarities and leads to item-based CF algorithms (cf. Chapter 2).

## 3.3 Projection-based Collaborative Filtering

Collaborative Filtering (CF for short) is one of the most used recommendation algorithms for personalized RS [3, 11]. Basically, it is an algorithm for matching people with similar interests under the assumption that similar people like similar things. Standard CF-based algorithms operate on second-order tensors (or matrices) representing binary relations between users and resources. In the following we describe how CF can be applied for computing tag and resource recommendations in STS through projection matrices.

### 3.3.1 Tag Recommendations

For computing tag recommendations based on the two projection matrices defined in Section 3.2, we first compute the neighborhood $N_u$ of a user $u$, by considering either the resources or the tags as components of the row vectors representing the user profiles. Having defined which projection matrix to use, we can now apply standard user-based CF (see Section 2.2). To compute user-user similarities the cosine similarity measure is typically used [5, 15], i. e.,

$$\mathrm{sim}(\mathbf{x}_u, \mathbf{x}_v) := \frac{\langle \mathbf{x}_u, \mathbf{x}_v \rangle}{\|\mathbf{x}_u\|\|\mathbf{x}_v\|} \tag{3.4}$$

The score of a tag $t \in T$ given a user $u \in U$ and a resource $r \in R$ as input is given by:

$$\hat{s}(u, r, t) := \sum_{v \in N_u} \mathrm{sim}(\mathbf{x}_u, \mathbf{x}_v)\delta\left((v, r, t) \in Y\right) \tag{3.5}$$

In words, the score of a tag $t \in T$ is the weighted sum of tag values amongst the best neighbors of the target user.

## 3.3.2 Tag-aware Recommendations

Note that if we use only the projection $\pi_{UR}Y$ for recommending resources, we would end up at the standard user-based (or item-based) CF algorithms. However, tags provide additional information about the user preferences, and thus can be exploited to boost the recommendation quality.

Firan et al. [1] suggested to first compute a ranked list of tags on the user-tag projection matrix $\pi_{UT}Y$, whereby the resources annotated with these tags are aggregated, ranked, and finally presented to the target user. But by using only $\pi_{UT}Y$ for recommending resources, one discards the information on the preferences of the target users for resources, which in this case, is the key mode of interest.

Tso-Sutter et al. [15] proposed to handle this issue by extending the typical user-resource matrix with tags as pseudo users and pseudo resources (see Figure 3.2). Note that in this way, the user (or resource) profile is automatically enriched with tags. A fusion algorithm is then proposed for combining the user-based ($ucf$) and item-based CF ($icf$) predictions over the extended matrix. For a given target user $u \in U$, the score of resource $r \in R$ is computed by

$$\hat{s}^{\mathrm{iucf}}(u, r) := \lambda \cdot \hat{s}^{\mathrm{ucf}}(u, r) + (1 - \lambda) \cdot \hat{s}^{\mathrm{icf}}(u, r) \tag{3.6}$$

where $\hat{s}^{\mathrm{ucf}}(u, r)$ and $\hat{s}^{\mathrm{icf}}(u, r)$ are the individual scores computed by user-based and item-based CF respectively, and $\lambda$ is a parameter controlling the influences of these individual scores.

Still based on projections of the original set $Y$ of ternary relations, Wetzker et al. [16] combined a probabilistic latent semantic analysis (PLSA) model [4] with tags for the recommendation of resources. In the standard PLSA, the probability that a resource co-occurs with a given user can be computed by

$$P(r \mid u) := \sum_{z \in Z} P(r \mid z)P(z \mid u), \tag{3.7}$$

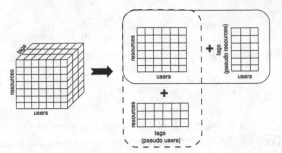

**Fig. 3.2** Extending the user-resource matrix horizontally by including tags as pseudo resources and vertically by including tags as pseudo users. Figure is adapted from Tso-Sutter et al. [15].

where $z$ is a hidden topic variable and is assumed to be the origin of observed co-occurrence distributions between users and resources. The same hidden topics are then assumed to be the origin of resource/tag co-occurrences, i. e.,

$$P(r \mid t) := \sum_{z \in Z} P(r \mid z) P(z \mid t). \tag{3.8}$$

Both models are then combined on the common factor $P(r \mid z)$ by maximizing the log-likelihood function

$$L := \sum_{r \in R} \left[ \lambda \sum_{u \in U} |Y_{u,r}| \log P(r \mid u) + (1 - \lambda) \sum_{t \in T} |Y_{r,t}| \log P(r \mid t) \right], \tag{3.9}$$

where $\lambda$ is a predefined weight balancing the influence of each model. The usual Expectation-Maximization (EM) algorithm is then applied for performing maximum likelihood estimation for the model. Resources for a given user $u$ are then weighted by the probability $P(r \mid u)$ (see Equation 3.7), ranked, and the top ranked resources are finally recommended.

### 3.3.3 User Recommendations

For recommending users, one can either compute the neighborhood of the target user based on $\pi_{UT} Y$ or $\pi_{UR} Y$ and recommend the $k$-best neighbors. In order to compute a neighborhood that takes into account both $\pi_{UT} Y$ or $\pi_{UR} Y$ at the same time, one could, for example, either recommend the users in the neighborhood computed on the matrix extensions proposed by [15] (see Figure 3.2) or compute a linear combination of the user similarities based on the user-resource and user-tag projection matrices.

## 3.3.4 Remarks on Complexity

Neighborhood-based CF usually suffers from scalability problems, given that
the whole input matrix needs to be kept in memory. In STS, one may have to
eventually keep more than one matrix in memory, depending on which kind
of projections one wants to operate upon. To compute recommendations we
usually need three steps:

1. Computation of projections: In order to compose the projections, we need
   to determine the $(u, r), (u, t)$ and/or $(r, t)$ co-occurrences. For that, we just
   need to do a linear scan in $Y$.
2. Neighborhood computation: In traditional user-based CF algorithms, the
   computation of the neighborhood $N_u$ is usually linear in the number of
   users, as one needs to compute the similarity of a given test user with all
   the other users in the database. In addition, we need to sort the similarities
   in order to determine the $k$-nearest neighbors.
3. Recommendations: For predicting the top-$n$ tag/resource recommendations for a given test user, we need to: (i) count the tags/resources co-
   occurrences with the nearest neighbors $N_u$, (ii) weigh each co-occurrence
   by the corresponding neighbor similarity, and (iii) sort the tags/resources
   based on their weights (e. g., Equation 3.5).

## 3.4 Further Reading

Delicious was one of the first STS to announce tag and resource recommendation services.[1] Although, no algorithmic details were published, we assume
that these recommendations are based on counting, like those presented in
Section 3.1. Counting-based methods were first used as baselines in [8, 6]
in the context of tag recommendations. Jäschke et al. [5] presented ensembles of counting-based methods and showed empirically that they perform
surprisingly well in comparison to more complex methods.

AutoTag [9] was one of the first tools designed to suggest tags for weblog
posts using collaborative filtering methods. AutoTag identifies useful tags for
a post by examining tags assigned to similar posts. Once the user supplies
a weblog post, posts which are similar to it are identified. Next, the tags
assigned to these posts are aggregated, creating a ranked list of likely tags to
recommend. AutoTag filters and reranks this tag list; finally, the top-ranked
tags are presented to the user, who can now select the tags to annotate to
the post of interest.

TagAssist [14] is another system that also provides tag suggestions for new
blog posts by using existing tagged posts. The system is able to increase the
quality of suggested tags by performing lossless compression over existing tag

---

[1] http://blog.delicious.com/blog/2005/08/people_who_like.html

data. TagAssist outperforms AutoTag in terms of accuracy, by introducing tag compression and case evaluation to filter and rank tag suggestions.

Peng et al. [10] proposed a method of joint Resource-Tag Recommendations. In particular, their method first generates joint resource-tag recommendations, with tags indicating topical interests of users in target resources. These joint recommendations are then refined by the "wisdom of the crowd" and projected to the resource space for final resource recommendations. Peng et al.'s [10] approach also integrates two different research directions. It combines the models of recent studies that try to represent the ternary (i. e., user, resource, tag) relationship as tensors with other models which rely on the bipartite interactions between any two of these three entities. Moreover, they have experimentally shown that their method outperforms the PLSA model proposed by Wetzker et al. [16].

Another research work that provides resources recommendations based on tags are Tagommenders [13]. Tagommenders are recommender algorithms that predict users' preferences for resources based on their inferred preferences for tags. The users' preferences can be inferred using tag signals, e. g., the tags he/she selected for browsing resources or the tags he used to annotate his/her resources, or resource signals e. g., a movie rating or a click in movie hyperlink. Based on these inferred resource/tag preferences, the authors proposed to combine tag preference inference algorithms with tag-aware recommenders and showed empirically that their approach outperforms classic CF algorithms and may lead to novel interfaces for recommender systems.

Santander and Brusilovsky [12] proposed several collaborative filtering techniques for the recommendation of scientific articles. In particular, they developed and compared four collaborative filtering approaches for resource recommendation in CiteULike. The first one and the baseline was the classic CF (CCF). The second approach, Neighbor-weighted Collaborative Filtering (NwCF), enhances traditional ranking (prediction) by taking into account the number of raters in the ranking formula of the recommendations. The third approach explores an innovative way to form the user neighborhood based on the Okapi BM25 [7] model over tags, while keeping the CCF ranking step intact. Okapi BM25 refers to a ranking function used by search engines to rank matching documents according to their relevance to a given search query. Finally, the combination of Okapi BM25 with NwCF approach uses the Okapi BM25 for neighborhood formation and NwCF for ranking. Their results demonstrate that the hybrid BM25-NwCF approach gives the best results, by combining the potential of each approach, i. e., increasing the coverage of users and items with BM25 and improving the precision with NwCF.

Finally, Gemmell et al. [2] proposed a weighted hybrid tag recommender that blends multiple recommendation components drawing separately on complementary dimensions. In particular, the weighted hybrid tag recommender consists of two popularity/counting-based recommenders and four collaborative filtering recommenders ($KNN_{ur}$, $KNN_{ut}$, $KNN_{ru}$, $KNN_{rt}$).

All different recommenders scores are combined in a linear model, which assigns different weights (importance) to each recommender.

# References

1. Claudiu S. Firan, Wolfgang Nejdl, and Raluca Paiu. The benefit of using tag-based profiles. In *LA-WEB '07: Proceedings of the 2007 Latin American Web Conference*, pages 32–41, Washington, DC, USA, 2007. IEEE Computer Society.

2. Jonathan Gemmell, Thomas Schimoler, Bamshad Mobasher, and Robin Burke. Hybrid tag recommendation for social annotation systems. In *Proceedings of the 19th ACM international conference on Information and knowledge management*, CIKM '10, pages 829–838, New York, NY, USA, 2010. ACM.

3. Jon Herlocker, Joseph A. Konstan, and John Riedl. An empirical analysis of design choices in neighborhood-based collaborative filtering algorithms. *Inf. Retr.*, 5:287–310, October 2002.

4. Thomas Hofmann. Probabilistic latent semantic indexing. In *SIGIR '99: Proceedings of the 22nd annual international ACM SIGIR conference on Research and development in information retrieval*, pages 50–57, New York, NY, USA, 1999. ACM.

5. Robert Jäschke, Leandro Marinho, Andreas Hotho, Lars Schmidt-Thieme, and Gerd Stumme. Tag recommendations in social bookmarking systems. *AI Communications*, 21(4):231–247, 2008.

6. Robert Jäschke, Leandro Balby Marinho, Andreas Hotho, Lars Schmidt-Thieme, and Gerd Stumme. Tag recommendations in folksonomies. In Joost N. Kok, Jacek Koronacki, Ramon López de Mántaras, Stan Matwin, Dunja Mladenic, and Andrzej Skowron, editors, *Knowledge Discovery in Databases: PKDD 2007, 11th European Conference on Principles and Practice of Knowledge Discovery in Databases*, volume 4702 of *Lecture Notes in Computer Science*, pages 506–514, Berlin, Heidelberg, 2007. Springer.

7. Hinrich Schütze Christopher D. Manning. *Introduction to Information Retrieval*. ISBN 978-0-521-86571-5. Cambridge University Press, May 2008.

8. Leandro Balby Marinho and Lars Schmidt-Thieme. Collaborative tag recommendations. In *GFKL '07: Proceedings of the 31st Annual Conference of the Gesellschaft für Klassifikation (GfKl), Freiburg*, pages 533–540. Springer, 2007.

9. Gilad Mishne. Autotag: A collaborative approach to automated tag assignment for weblog posts. In *WWW '06: Proceedings of the 15th International Conference on World Wide Web*, pages 953–954, New York, NY, USA, 2006. ACM Press.

10. Jing Peng, Daniel Dajun Zeng, Huimin Zhao, and Fei-yue Wang. Collaborative filtering in social tagging systems based on joint item-tag recommendations. In *Proceedings of the 19th ACM international conference on Information and knowledge management*, CIKM '10, pages 809–818, New York, NY, USA, 2010. ACM.
11. Paul Resnick, Neophytos Iacovou, Mitesh Suchak, Peter Bergstrom, and John Riedl. Grouplens: an open architecture for collaborative filtering of netnews. In *CSCW '94: Proceedings of the 1994 ACM conference on Computer supported cooperative work*, pages 175–186, New York, NY, USA, 1994. ACM.
12. Denis P. Santander and Peter Brusilovsky. Improving Collaborative Filtering in Social Tagging Systems for the Recommendation of Scientific Articles. *Web Intelligence and Intelligent Agent Technology, IEEE/WIC/ACM International Conference on*, 1:136–142, 2010.
13. Shilad Sen, Jesse Vig, and John Riedl. Tagommenders: connecting users to items through tags. In *Proceedings of the 18th international conference on World wide web*, WWW '09, pages 671–680, New York, NY, USA, 2009. ACM.
14. Sanjay Sood, Sara Owsley, Kristian Hammond, and Larry Birnbaum. Tagassist: Automatic tag suggestion for blog posts. In *ICWSM '07: Proceedings of the International Conference on Weblogs and Social Media*, 2007.
15. Karen H. L. Tso-Sutter, Leandro Balby Marinho, and Lars Schmidt-Thieme. Tag-aware recommender systems by fusion of collaborative filtering algorithms. In *SAC '08: Proceedings of the 2008 ACM symposium on Applied computing*, pages 1995–1999. ACM, 2008.
16. Robert Wetzker, Winfried Umbrath, and Alan Said. A hybrid approach to item recommendation in folksonomies. In *ESAIR '09: Proceedings of the WSDM '09 Workshop on Exploiting Semantic Annotations in Information Retrieval*, pages 25–29. ACM, 2009.

# Chapter 4
# Advanced Techniques

In this chapter we describe the state-of-the-art in social tagging recommender systems. Many of the algorithms presented here borrow ideas and techniques from other areas such as information retrieval, machine learning, and statistical relational learning. In Section 4.3 we also describe many approaches for exploiting additional sources of information such as the content of resources and the social relations of users.

## 4.1 Factorization Models

The idea of representing data in lower dimensional spaces has been extensively used in natural language processing where it is usually known as *latent semantic analysis* (LSA), or *latent semantic indexing* (LSI) in the particular context of information retrieval [9]. LSA-based methods are appealing for scenarios in which the data is too large, too sparse, and/or too noisy, since the reduced representation of the data can be interpreted as a de-noisified approximation of the "true" data. Given that these problems, namely large-scale, noise, and sparsity, are recurrent issues in recommender systems, LSA-based techniques appear as an interesting tool to be exploited. Hofmann [15], for example, used a probabilistic version of LSI for the prediction of resources/ratings in recommender systems. There, and in any low dimensional factor model in fact, it is assumed that there is only a small number of factors influencing the users' preferences, and that a user's preference for a resource is determined by how each factor applies to the user and the resource. More recently, due to the Netflix challenge[1], research on matrix factorization methods, a class of latent factor models, gained renewed momentum in the recommender systems

---

[1] The Netflix challenge was a competition for the best recommender system algorithm to predict user ratings for movies. The competition was held by Netflix (http://www.netflixprize.com/), an on-line DVD-rental service.

literature, given that many of the best performing methods on the challenge were based on matrix factorization techniques [26, 45, 27].

As mentioned in Chapter 1, the ternary relation $Y$ can be represented as a third-order tensor $\mathcal{A}$, such that tensor factorization techniques can be employed in order to exploit the underlying latent semantic structure in $\mathcal{A}$. While the idea of computing low rank tensor approximations has already been used for many different purposes [28, 46, 25, 51, 7, 47], just recently it has been applied for the problem of recommendations in STS. The basic idea is to cast the recommendation problem as a third-order tensor completion problem — completing the non-observed entries in $\mathcal{A}$. The approximation $\hat{\mathcal{A}}$ of tensor $\mathcal{A}$ yields the recommendation scores, e.g., for tag prediction:

$$\hat{s}(u, r, t) := \hat{a}_{u,r,t}$$

In the following we present several approaches for recommending in STS based on tensor factorization.

## 4.1.1 Higher Order Singular Value Decomposition – HOSVD on Tensors

HOSVD is a generalization of singular value decomposition and has been successfully applied in several areas. In this section, we summarize the HOSVD procedure.

### 4.1.1.1 From SVD to HOSVD

The singular value decomposition (SVD) [1] of a matrix $F_{I_1 \times I_2}$ can be written as a product of three matrices, as shown in Equation 4.1:

$$F_{I_1 \times I_2} = Q_{I_1 \times I_1} \cdot S_{I_1 \times I_2} \cdot V^T_{I_2 \times I_2,} \tag{4.1}$$

where $Q$ is the matrix with the left singular vectors of $F$, $V^T$ is the transpose of the matrix $V$ with the right singular vectors of $F$, and $S$ is the diagonal matrix of ordered singular values [2] of $F$.

By preserving only the largest $k < \min\{I_1, I_2\}$ singular values of $S$, SVD results in matrix $\hat{F}$, which is an approximation of $F$. In information retrieval, this technique is used by LSI [11], to deal with the latent semantic associations of terms in texts and to reveal the major trends in $F$.

Formally, a *tensor* is a multi-dimensional matrix. A $N$-order tensor $\mathcal{A}$ is denoted as $\mathcal{A} \in \mathbb{R}^{I_1 \cdots I_N}$, with elements $a_{i_1, \ldots, i_N}$. The high-order singular

---

[2] The singular values determined by the factorization of Equation 4.1 are unique and satisfy $\sigma_1 \geq \sigma_2 \geq \sigma_3 \geq \cdots \geq \sigma_{I_2} \geq 0$.

value decomposition [28] generalizes the SVD computation to tensors. To apply HOSVD on a 3-order tensor $\mathcal{A}$, three *matrix unfolding*[3] operations are defined as follows [28]:

$$A_1 \in \mathbb{R}^{I_1 \times (I_2 I_3)}, \qquad A_2 \in \mathbb{R}^{I_2 \times (I_1 I_3)}, \qquad A_3 \in \mathbb{R}^{(I_1 I_2) \times I_3}$$

where $A_1, A_2, A_3$ are called the mode-1, mode-2, mode-3 matrix unfolding of $\mathcal{A}$, respectively. The unfoldings of $\mathcal{A}$ in the three modes are illustrated in Figure 4.1.

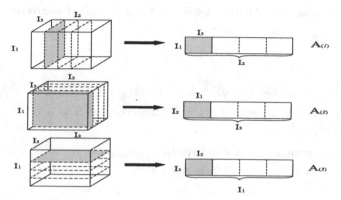

**Fig. 4.1** Visualization of the three unfoldings of a 3-order tensor.

**Example 1:** Define a tensor $\mathcal{A} \in \mathbb{R}^{3 \times 2 \times 3}$ by $a_{1,1,1} = a_{1,1,2} = a_{2,1,1} = -a_{2,1,2} = 1, a_{2,1,3} = a_{3,1,1} = a_{3,1,3} = a_{1,2,1} = a_{1,2,2} = a_{2,2,1} = -a_{2,2,2} = 2, a_{2,2,3} = a_{3,2,1} = a_{3,2,3} = 4, a_{1,1,3} = a_{3,1,2} = a_{1,2,3} = a_{3,2,2} = 0$. The tensor and its mode-1 matrix unfolding $A_1 \in \mathbb{R}^{I_1 \times I_2 I_3}$ are illustrated in Figure 4.2.

Next, we define the mode-$n$ product of a $N$-order tensor $\mathcal{A} \in \mathbb{R}^{I_1 \times \cdots \times I_N}$ by a matrix $Q \in \mathbb{R}^{J_n \times I_n}$, which is denoted as $\mathcal{A} \times_n Q$. The result of the mode-$n$ product is an $(I_1 \times I_2 \times \cdots \times I_{n-1} \times J_n \times I_{n+1} \times \cdots \times I_N)$-tensor, the entries of which are defined as follows:

$$(\mathcal{A} \times_n Q)_{i_1 i_2 \ldots i_{n-1} j_n i_{n+1} \ldots i_N} = \sum_{i_n} a_{i_1 i_2 \ldots i_{n-1} i_n i_{n+1} \ldots i_N} q_{j_n, i_n} \qquad (4.2)$$

Since we focus on 3-order tensors, $n \in \{1, 2, 3\}$, we use mode-1, mode-2, and mode-3 products.

In terms of mode-$n$ products, SVD on a regular two-dimensional matrix (i.e., 2-order tensor), can be rewritten as follows [28]:

---

[3] We define as "matrix unfolding" of a given tensor the matrix representations of that tensor in which all the column (row, ... ) vectors are stacked one after the other.

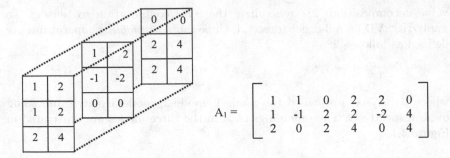

**Fig. 4.2** Visualization of tensor $\mathcal{A} \in \mathbb{R}^{3 \times 2 \times 3}$ and its mode-1 matrix unfolding.

$$F = S \times_1 Q^{(1)} \times_2 Q^{(2)} \tag{4.3}$$

where $U^{(1)} = (q_1^{(1)} q_2^{(1)} \ldots q_{I_1}^{(1)})$ is a *unitary* $(I_1 \times I_1)$-matrix [4], $Q^{(2)} = (q_1^{(2)} q_2^{(2)} \ldots q_{I_1}^{(2)})$ is a *unitary* $(I_2 \times I_2)$-matrix, and $S$ is a $(I_1 \times I_2)$-matrix with the properties of:

i. pseudo-diagonality: $S = \mathrm{diag}(\sigma_1, \sigma_2, \ldots, \sigma_{\min\{I_1, I_2\}})$
ii. ordering: $\sigma_1 \geq \sigma_2 \geq \cdots \geq \sigma_{\min\{I_1, I_2\}} \geq 0$.

By extending this form of SVD, HOSVD of a 3-order tensor $\mathcal{A}$ can be written as follows [28]:

$$\mathcal{A} = S \times_1 Q^{(1)} \times_2 Q^{(2)} \times_3 Q^{(3)} \tag{4.4}$$

where $Q^{(1)}$, $Q^{(2)}$, $Q^{(3)}$ contain the orthonormal vectors (called the mode-1, mode-2 and mode-3 singular vectors, respectively) spanning the column space of the $A_1, A_2, A_3$ matrix unfoldings. $S$ is called core tensor and has the property of "all orthogonality".[5] This decomposition also refers to a general factorization model known as Tucker decomposition [49].

---

[4] An $n \times n$ matrix $Q$ is said to be unitary if its column vectors form an orthonormal set in the complex inner product space $\mathbb{C}^n$. That is, $Q \times Q^T = I_n$.

[5] All- orthogonality means that the different "horizontal matrices" of $S$ (the first index $i_1$ is kept fixed, while the two other indices, $i_2$ and $i_3$, are free) are mutually orthogonal with respect to the scalar product of matrices (i. e., the sum of the products of the corresponding entries vanishes); at the same time, the different "frontal" matrices ($i_2$ fixed) and the different "vertical" matrices ($i_3$ fixed) should be mutually orthogonal as well. For more information, see [28].

### 4.1.1.2 HOSVD for Recommendations in STS

In this subsection we elaborate on how HOSVD can be employed for computing recommendations in STS and present an example on how one can recommend resources according to the detected latent associations. Although we illustrate only the recommendation of resources, once the approximation $\hat{\mathcal{A}}$ is computed the recommendation of users or tags is straightforward [48].

Recall from Chapter 3 that $Y$ can be represented by the binary tensor $\mathcal{A} = (a_{u,r,t}) \in \mathbb{R}^{|U| \times |R| \times |T|}$ where 1 indicates observed tag assignments and 0 missing values, i.e.,

$$a_{u,r,t} := \begin{cases} 1, & (u,r,t) \in Y \\ 0, & \text{else} \end{cases}$$

Now, we express the tensor decomposition as

$$\hat{\mathcal{A}} := \hat{C} \times_u \hat{U} \times_r \hat{R} \times_t \hat{T} \tag{4.5}$$

where $\hat{U}$, $\hat{R}$, and $\hat{T}$ are low-rank feature matrices representing a mode, i.e., user, resources, and tags respectively, in terms of its small number of latent dimensions $k_U$, $k_R$, $k_T$, and $\hat{C} \in \mathbb{R}^{k_U \times k_R \times k_T}$ is the core tensor representing interactions between the latent factors. The model parameters to be optimized are represented by the quadruple $\hat{\theta} := (\hat{C}, \hat{U}, \hat{R}, \hat{T})$ (see Figure 4.3).

The basic idea of the HOSVD algorithm is to minimize an element-wise loss on the elements of $\hat{\mathcal{A}}$ by optimizing the square loss, i.e.,

$$\underset{\hat{\theta}}{\text{argmin}} \sum_{(u,r,t) \in Y} (\hat{a}_{u,r,t} - a_{u,r,t})^2$$

After the parameters are optimized, predictions can be done as follows:

$$\hat{s}(u,r,t) := \sum_{\tilde{u}} \sum_{\tilde{r}} \sum_{\tilde{t}} \hat{c}_{\tilde{u},\tilde{r},\tilde{t}} \cdot \hat{u}_{u,\tilde{u}} \cdot \hat{r}_{r,\tilde{r}} \cdot \hat{t}_{t,\tilde{t}} \tag{4.6}$$

where indices over the feature dimension of a feature matrix are marked with a tilde, and elements of a feature matrix are marked with a hat (e.g., $\hat{t}_{t,\tilde{t}}$).

**Example 2:** The HOSVD algorithm takes $\mathcal{A}$ as input and outputs the reconstructed tensor $\hat{\mathcal{A}}$. $\hat{\mathcal{A}}$ measures the strength of associations between users, resources, and tags. Each element of $\hat{\mathcal{A}}$ can be represented by a quadruplet $\{u,r,t,p\}$, where $p$ measures the likeliness that user $u$ will tag resource $r$ with tag $t$. Therefore, resources can be recommended to $u$ according to their weights associated with the $\{u,t\}$ pair.

In this subsection, in order to illustrate how HOSVD for resource recommendation works, we apply HOSVD to a toy example. As illustrated in Figure 4.4, three users tagged three different resources (web links). In Figure 4.4,

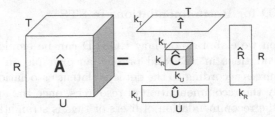

**Fig. 4.3** Tensor decomposition in STS. Figure adapted from [43].

the part of an arrow line (sequence of arrows with the same annotation) between a user and a resource represents that the user tagged the corresponding resource, and the part between a resource and a tag indicates that the user tagged this resource with the corresponding tag. Thus, the annotated numbers on the arrow lines gives the correspondence between the three types of objects. For example, user $u_1$ tagged resource $r_1$ with tag "BMW", denoted as $t_1$. The remaining tags are "Jaguar", denoted as $t_2$, "CAT", denoted as $t_3$.

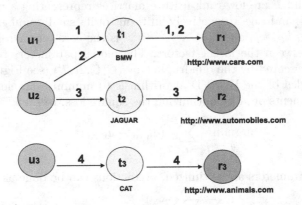

**Fig. 4.4** Usage data of the running example.

From Figure 4.4, we can see that users $u_1$ and $u_2$ have common interests on cars, while user $u_3$ is interested in cats. A 3-order tensor $\mathcal{A} \in \mathbb{R}^{3 \times 3 \times 3}$, can be constructed from the usage data. We use the co-occurrence frequency (denoted as weights) of each triplet user, resource, and tag as the elements of tensor $\mathcal{A}$, which are given in Table 4.1. Note that all associated weights are initialized to 1. Figure 4.5 presents the tensor construction of our running example.

After performing the tensor reduction analysis, we can get the reconstructed tensor of $\hat{\mathcal{A}}$, which is presented in Table 4.2, whereas Figure 4.6 depicts the contents of $\hat{\mathcal{A}}$ graphically (the weights are omitted). As shown in

**Table 4.1** Tensor constructed from the usage data of the running example.

| Arrow Line | User | Resource | Tag | Weight |
|:---:|:---:|:---:|:---:|:---:|
| 1 | $u_1$ | $r_1$ | $t_1$ | 1 |
| 2 | $u_2$ | $r_1$ | $t_1$ | 1 |
| 3 | $u_2$ | $r_2$ | $t_2$ | 1 |
| 4 | $u_3$ | $r_3$ | $t_3$ | 1 |

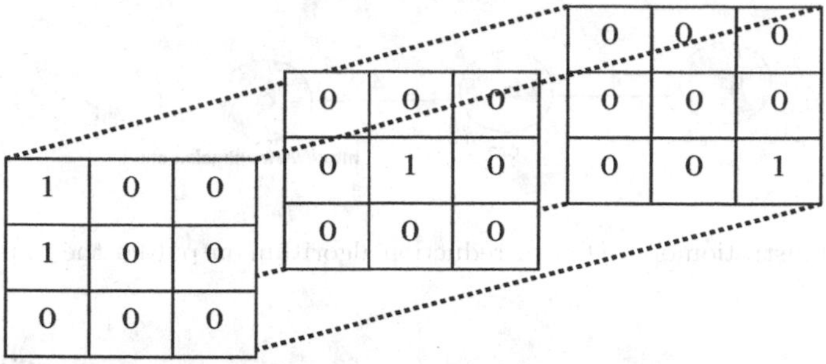

**Fig. 4.5** The tensor construction of our running example.

Table 4.2 and Figure 4.6, the output of the tensor reduction algorithm for the running example is interesting, because a new association among these objects is revealed. The new association is between $u_1$, $r_2$, and $t_2$. This association is represented with the last (bold faced) row in Table 4.2 and with the dashed arrow line in Figure 4.6).

If we have to recommend to $u_1$ a resource for tag $t_2$, then there is no direct indication for this task in the original tensor $\mathcal{A}$. However, we see that in Table 4.2 the element of $\hat{\mathcal{A}}$ associated with $(u_1, t_2, r_2)$ is 0.44, whereas for $u_1$ there is no other element associating other tags with $r_2$. Thus, we recommend resource $r_2$ to user $u_1$, who used tag $t_2$. For the current example, the resulting $\hat{\mathcal{A}}$ tensor is presented in Figure 4.7.

**Table 4.2** Tensor constructed from the usage data of the running example.

| Arrow Line | User | Item | Tag | Weight |
|:---:|:---:|:---:|:---:|:---:|
| 1 | $u_1$ | $r_1$ | $t_1$ | 0.72 |
| 2 | $u_2$ | $r_1$ | $t_1$ | 1.17 |
| 3 | $u_2$ | $r_2$ | $t_2$ | 0.72 |
| 4 | $u_3$ | $r_3$ | $t_3$ | 1 |
| **5** | **$u_1$** | **$r_2$** | **$t_2$** | **0.44** |

The resulting recommendation is reasonable, because $u_1$ is interested in cars rather than cats. That is, the tensor reduction approach is able to capture the latent associations among the multi-type data objects: user, resources, and tags. The associations can then be used to improve the resource recommendation procedure.

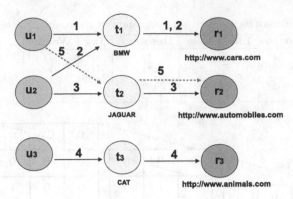

**Fig. 4.6** Illustration of the tensor reduction algorithm output for the running example.

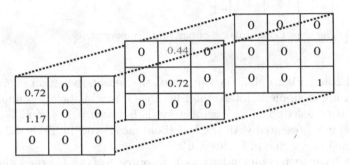

**Fig. 4.7** The resulting $\hat{A}$ tensor for the running example.

### 4.1.1.3 Combining HOSVD with Content-based Methods

Social tagging has become increasingly popular in music information retrieval (MIR). It allows users to tag music resources like songs, albums, or artists. Social tags are valuable to MIR, because they comprise a multifaceted source of information about genre, style, mood, users' opinion, or instrumentation.

Symeonidis et al. [37] examined the problem of personalized song recommendation (i. e., resource recommendation) based on social tags. They proposed the modeling of social tagging data with 3-order tensors, which capture cubic (3-way) correlations between users-tags-music items. The discovery of latent structure in this model is performed with HOSVD, which helps to provide accurate and personalized recommendations, i. e., adapted to the particular users' preferences.

However, the aforementioned model suffers from sparsity that incurs in social tagging data. Thus, to further improve the quality of recommendation, Nanopoulos et al. [36] enhanced the HOSVD model with a tag-propagation scheme that uses similarity values computed between the music resources based on audio features. As a result, this hybrid model effectively combines both information about social tags and audio features. Nanopoulos et al. [36] examined experimentally the performance of the proposed method with real data from Last.fm. Their results indicate the superiority of the proposed approach compared to existing methods that suppress the cubic relationships that are inherent in social tagging data. Additionally, their results suggest that the combination of social tagging data with audio features is preferable than the use of the former alone.

#### 4.1.1.4 Limitations of HOSVD

The HOSVD approach has two major drawbacks:

Modeling:   The runtime complexity is cubic in the size of the latent dimensions. This can be seen in Equation 4.6, where three nested sums have to be calculated just for predicting a single (user, resource, tag)-triple. There are several approaches to improve the efficiency of HOSVD [25, 50, 10].

Learning:   HOSVD is optimized for least-squares on the whole tensor $\mathcal{A}$. However, recommendation is a ranking task not a regression task and also the non-observed posts are not taken into account by HOSVD.

We will study both issues in the following.

## 4.1.2 Scalable Factorization Models

The limitation in runtime of HOSVD stems from its model which is the Tucker Decomposition. In the following, we will discuss several factorization models that have been proposed for tag recommendation. We investigate their model assumptions, complexity and their relations among each other.

#### 4.1.2.1 Tucker Decomposition

The underlying tensor factorization model of HOSVD is the Tucker Decomposition (TD) [49]. As noted before, for tag recommendation, the model reads:

$$\hat{\mathcal{A}} := \hat{C} \times_u \hat{U} \times_r \hat{R} \times_t \hat{T} \tag{4.7}$$

or equivalently

$$\hat{a}_{u,r,t} = \sum_{\tilde{u}=1}^{k_U} \sum_{\tilde{r}=1}^{k_R} \sum_{\tilde{t}=1}^{k_T} \hat{c}_{\tilde{u},\tilde{r},\tilde{t}} \cdot \hat{u}_{u,\tilde{u}} \cdot \hat{r}_{r,\tilde{r}} \cdot \hat{t}_{t,\tilde{t}} \tag{4.8}$$

The reason for the cubic complexity (i. e., $O(k^3)$ with $k := \min(k_U, k_R, k_T)$) of TD is the core tensor.

### 4.1.2.2 Parallel Factor Analysis (PARAFAC)

The Parallel Factor Analysis (PARAFAC) [14] model aka canonical decomposition [5] reduces the complexity of the TD model by assuming a diagonal core tensor.

$$c_{\tilde{u},\tilde{r},\tilde{t}} \overset{!}{=} \begin{cases} 1, & \text{if } \tilde{u} = \tilde{r} = \tilde{t} \\ 0, & \text{else} \end{cases} \tag{4.9}$$

which allows to rewrite the model equation:

$$\hat{a}_{u,r,t} = \sum_{f=1}^{k} \hat{u}_{u,f} \cdot \hat{r}_{r,f} \cdot \hat{t}_{t,f} \tag{4.10}$$

In contrast to TD, the model equation of PARAFAC can be computed in $O(k)$. In total, the model parameters $\hat{\theta}$ of the PARAFAC model are:

$$\hat{U} \in \mathbb{R}^{|U| \times k}, \quad \hat{R} \in \mathbb{R}^{|R| \times k}, \quad \hat{T} \in \mathbb{R}^{|T| \times k} \tag{4.11}$$

The assumption of a diagonal core tensor is a restriction of the TD model.

### 4.1.2.3 Pairwise Interaction Tensor Factorization (PITF)

Whereas TD and PARAFAC directly express a ternary relation, the idea of the pairwise interaction tensor factorization (PITF) [44] is to model pairwise interactions instead. The motivation is that observations are typically very limited and sparse in tag recommendation data, and thus it is often easier to estimate pairwise interactions than ternary ones. This assumption is reflected in the model equation of PITF which reads:

$$\hat{a}_{u,r,t} = \sum_{f}^{k} \hat{u}_{u,f} \cdot \hat{t}^{U}_{t,f} + \sum_{f}^{k} \hat{r}_{r,f} \cdot \hat{t}^{R}_{t,f} \tag{4.12}$$

with model parameters $\hat{\theta}$

$$\hat{U} \in \mathbb{R}^{|U| \times k}, \quad \hat{R} \in \mathbb{R}^{|R| \times k}, \quad \hat{T^U} \in \mathbb{R}^{|T| \times k}, \quad \hat{T^R} \in \mathbb{R}^{|T| \times k} \tag{4.13}$$

Note that in contrast to PARAFAC, there are two factor matrices for the tags. One $(T^U)$ for the interaction of tags with users and a second one $(T^R)$ for the interaction of tags with resources. In general, one could also add an interaction between users and resources but it would not play any role for tag recommendation because it is invariant to ranking [44] if a proper ranking optimization is chosen (see Sec. 4.1.3).

### 4.1.2.4 Factorization Machines (FM)

The tensor factorization approaches discussed so far can only deal with prediction problems involving three categorical variables (i.e., user, resource and tag). Often additional knowledge is present, e.g., the title or text of resources or some information about the user or context-information of the tagging event such as time.

Factorization machines (FM) [41] have been proposed as a generic model that allows to describe a wide variety of data by feature engineering. FMs combine the flexibility of feature engineering with the advantages of factorization. The model equation of an FM reads

$$\hat{s}(\mathbf{x}) = w_0 + \sum_{j=1}^{p} x_j \, w_j + \sum_{j=1}^{p} \sum_{j'>j}^{p} x_j \, x_{j'} \sum_{f=1}^{k} v_{j,f} \, v_{j',f} \qquad (4.14)$$

where $\mathbf{x}$ is the input data describing e.g., a (user, resource, tag)-triple and $\hat{\theta}$ are the model parameters

$$w_0 \in \mathbb{R}, \quad \mathbf{w} \in \mathbb{R}^p, \quad V \in \mathbb{R}^{p \times k}. \qquad (4.15)$$

For applying FMs to social tag recommendation, the prediction $\hat{s}$ of an FM gives the score for a (user, resource, tag)-triple which is encoded using the real-valued feature vector $\mathbf{x}$. In general, any information that can be described with real-valued features can be used in $\mathbf{x}$. A simple encoding for tag recommendation has been proposed in [41], where the triple about user $u$, resource $r$, tag $t$ is described with a feature vector $\mathbf{x} \in \mathbb{R}^p$

$$\mathbf{x}_{u,r,t} := \Big( \underbrace{0, \ldots, \overset{u}{1}, 0, \ldots}_{|U|}, \underbrace{0, \ldots, \overset{r}{1}, 0, \ldots}_{|R|}, \underbrace{0, \ldots, \overset{t}{1}, 0, \ldots}_{|T|} \Big) \qquad (4.16)$$

where $p := |U| + |R| + |T|$. Using this encoding, the FM model is similar to PITF and generates also empirically a comparable quality in comparable runtime [41]. However, FMs are much more flexible as they allow to encode any kind of additional data, e.g., the input vector $\mathbf{x}$ can be extended with features for the words of a resource's title, the age of a user, the time, etc. which results in tag recommender that are content-aware, user-attribute-aware, time-aware, etc.

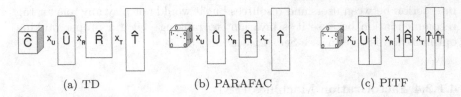

(a) TD                        (b) PARAFAC                        (c) PITF

**Fig. 4.8** Relationship between Tucker Decomposition, Parallel Factor Analysis (PARAFAC) and Pairwise Interaction Tensor Factorization (PITF) [40].

### 4.1.2.5 Relationship of Factorization Models

A graphical representation of Tucker Decomposition (TD), Parallel Factor Analysis (PARAFAC) and Pairwise Interaction Tensor Factorization (PITF) is shown in Figure 4.8. It can be seen that any PARAFAC model can be expressed by a TD model (with diagonal core tensor) and every PITF model can be expressed as a PARAFAC model where parts of PARAFAC's latent user (and resource) factors are constantly 1.

Let $\mathcal{M}$ be the set of models that can be represented by a model class. In [40] it is shown that for tag recommendation

$$\mathcal{M}^{\mathrm{TD}} \supset \mathcal{M}^{\mathrm{PARAFC}} \supset \mathcal{M}^{\mathrm{PITF}} \tag{4.17}$$

This means that any PARAFAC model can be expressed with a TD model but there are TD models that cannot be represented with a PARAFAC model. The same holds for PITF which is a true subclass of PARAFAC models. In [44, 40] it was pointed out that this does not mean that TD is guaranteed to have a higher prediction quality than PARAFAC and PARAFAC a higher quality than PITF. On the contrary, as all model parameters are estimated from limited data, restricting the expressiveness of a model can lead to higher prediction quality if the restriction is in line with the true parameters. Empirically this has been shown in [44] where PITF outperforms the more expressive TD and PARAFAC models in prediction quality. Rendle [40] mentions that the positive effect of restricting the expressiveness can also be achieved by regularization, e. g., assuming a non-zero mean for the prior distribution; e. g., non-zero Gaussian priors on some of PARAFACs factors would result in a PITF-like PARAFAC model. More details about the relationship of the factorization models as well as complexities can be found in [40].

## *4.1.3 Learning Tag Recommendation Models*

So far, factorization models have been discussed which rely on model parameters. Now, learning approaches are presented that find the optimal values for the model parameters $\hat{\theta}$ based on the observed data $Y$.

In the HOSVD approach, the TD model is optimized for least squares. However tag recommendation is a ranking rather than a regression task. In [43] and [44] it has been shown how model parameters can be optimized for optimal ranking of tags. First, the observed tag assignments are divided in positive, negative, and missing values as follows (cf. Section 1.4.1). Let $P_A$ be the set of all distinct user/resource combinations in $Y$ (called $\pi_{U,R}Y$ in Chapter 3), now the sets of positive and negative tags of a particular $(u, r) \in P_A$ are defined as

$$T_{u,r}^+ := \{t \mid (u,r) \in P_A \wedge (u,r,t) \in Y\}$$
$$T_{u,r}^- := \{t \mid (u,r) \in P_A \wedge (u,r,t) \notin Y\}$$

From this, pairwise tag ranking constraints can be defined for the values of $\hat{A}$:

$$a_{u,r,t_1} > a_{u,r,t_2} \Leftrightarrow (u,r,t_1) \in T_{u,r}^+ \wedge (u,r,t_2) \in T_{u,r}^- \qquad (4.18)$$

Two optimization criteria based on ranking constraints have been proposed: AUC optimization and pairwise classification.

**AUC Optimization.** In [43], the ranking statistic AUC (area under the ROC-curve) is maximized. The AUC measure for a particular post $(u, r) \in P_A$ is defined as:

$$\text{AUC}(\hat{\theta}, u, r) := \frac{1}{|T_{u,r}^+||T_{u,r}^-|} \sum_{t^+ \in T_{u,r}^+} \sum_{t^- \in T_{u,r}^-} H_{0.5}(\hat{a}_{u,r,t^+} - \hat{a}_{u,r,t^-}) \qquad (4.19)$$

where $H_\alpha$ is the Heaviside function:

$$H_\alpha := \begin{cases} 0, & x < 0 \\ \alpha, & x = 0 \\ 1, & x > 0 \end{cases} \qquad (4.20)$$

The overall optimization criterion with respect to the ranking statistic AUC and the observed data is then:

$$\text{AUC-OPT}(\hat{\theta}) := \sum_{(u,r) \in P_A} \text{AUC}(\hat{\theta}, u, r) \qquad (4.21)$$

Note that optimizing (4.19) directly is hard since $H_{0.5}$ is discontinuous and therefore not differentiable at 0. Hence, $H_\alpha$ is usually replaced by a smoother function that shares properties with the unit step function, for example, the s-shaped sigmoid function:

$$\sigma(x) := \frac{1}{1 + e^{-x}} \tag{4.22}$$

In [43], this approach of AUC optimization is named RTF (Ranking with Tensor Factorization) and applied to the TD model. For optimizing model parameters w. r. t. AUC-OPT (Equation 4.21), a post-wise gradient descent algorithm is proposed. Calculating the gradients per post $(u, r) \in P_{\mathcal{A}}$, allows caching parts of the TD model equation and speeds up learning (see [43] for details).

**Pairwise Classification.** The second optimization approach that has been proposed for parameter learning of tag recommenders is an adaptation of the Bayesian Personalized Ranking (BPR) [42] approach, which was originally used for item recommendation. BPR for tag recommendation [44] tries to find model parameters that satisfy as many ranking constraints (Equation 4.18) as possible. This can be formulated as a classification problem over ranking constraints. With a probabilistic treatment, the optimization criterion for maximum likelihood parameter estimation reads:

$$\text{BPR-OPT}(\hat{\theta}) := \sum_{(u,r) \in P_{\mathcal{A}}} \sum_{t^+ \in T_{u,r}^+} \sum_{t^- \in T_{u,r}^-} \ln \sigma(\hat{a}_{u,r,t^+} - \hat{a}_{u,r,t^-}) \tag{4.23}$$

Both AUC-OPT and BPR-OPT are closely related [42]. For optimizing model parameters w. r. t. BPR-OPT (Equation 4.23), a stochastic gradient descent learning method was proposed [44] where constraints (Equation 4.18, i. e., quadruples $(u, r, t_1, t_2)$) are drawn by bootstrapping.

The PITF model with BPR-optimization [44] was the winning approach of the ECML PKDD Discovery Challenge's task 2 on graph-based tag recommendations [16].

**Regularization.** Both AUC-OPT and BPR-OPT are typically extended with L2 regularization terms that favor small values for model parameters [43, 44]. L2-regularization is also known as Tikhonov regularization, ridge regression or Gaussian priors. The advantage of a regularized optimization criterion is that overfitting can be reduced. The regularized optimization task can be formalized as

$$\underset{\hat{\theta}}{\text{argmax}} \left( \text{OPT}(\hat{\theta}) + \lambda \, ||\hat{\theta}||^2 \right) \tag{4.24}$$

where $\lambda$ is the strength of the regularization. Often the model parameters are grouped and every group of model parameters has an individual regularization parameter $\lambda$.

## 4.2  Graph-based Models

In this section we present recommendation methods that exploit the hyper-graph representation of folksonomies, or transformations of it, for generating recommendations in social tagging systems.

### 4.2.1  PageRank-based Recommendations in STS

The analysis of hyperlinks and the graph structure of the Web has greatly contributed to the development of effective information retrieval techniques, turning web search engines into indispensable tools for any type of internet users. Link analysis for the Web borrows the ideas from citation analysis, in which citations represent the conferral of authority from a scholarly article to others [34]. In a similar way, link analysis on the Web considers hyperlinks from a web page to another as a conferral of authority.

The ranking algorithm PageRank [2], for example, reflects the idea that a web page is important if there are many pages linking to it, and if those pages are important themselves. Hotho et al. [17] employed the idea of PageRank for search and ranking in folksonomies. The key idea of the FolkRank algorithm is that a resource which is tagged with important tags by important users becomes important itself. Note that the same holds symmetrically for tags and users, which enables a global ranking of users, resources, and tags. Later on, Jäschke et al. [21] adapted the original FolkRank algorithm for personalized tag recommendations. In the following, we will first recall the principles of PageRank and then show how it can be used for computing recommendations in social tagging systems.

**PageRank.** PageRank employs the random surfer model, i.e., a random surfer begins at a web page and executes a random walk on the Web. So, at each time step, the surfer goes, usually with some uniform probability greater than 0, from his current web page A to another web page B that is linked by A. Proceeding in this way, the surfer will end up visiting some pages more often than others, with the most often visited pages reflecting pages with more links coming in from other frequently visited web pages. Hence, the basic idea is that pages visited more often are more important than others visited less frequently. Additionally, a *jump* operator is used to denote the probability that the user will "jump" to any other page in the Web, regardless whether it is linked by the current page or not. This operator reflects the situation where, for example, the user directly types a URL address into the URL bar of his browser, instead of following links from one page to another. In PageRank, the Web is usually represented as a graph $G = (V, E)$, in which $V$ is the set of known web pages and $E$ represents hyperlink relations between the web pages, e.g., $(x, y) \in E$ if there is a hyperlink from $x$ to $y$ or vice-versa.

One iteration of PageRank is computed as follows:

$$\mathbf{w}_{t+1} \leftarrow \lambda A^{\mathrm{T}} \mathbf{w}_t + (1 - \lambda) \mathbf{p} \tag{4.25}$$

where $\mathbf{w}$ is the surfer vector with one entry for each node in $V$, $A \in \mathbb{R}^{|V| \times |V|}$ is the row-stochastic version of the adjacency matrix $A'$ associated with the web graph[6], that is, $A_{i,j} = \frac{A'_{i,j}}{\deg(i)}$ if $\{i,j\} \in E$ and 0 else, $\mathbf{p}$ is a vector that can be used for asserting preferences for specific nodes, and the jumping factor $0 < \lambda < 1$ is used for determining the strength of the influence of $\mathbf{p}$. By normalization of the vector $\mathbf{p}$, the equality $||\mathbf{w}||_1 = ||\mathbf{p}||_1$ is enforced. This ensures that the weight in the system will remain constant if each page has at least one out-link. Therefore one usually removes all pages without outgoing links before applying PageRank. The random walk with the jump operator induces an ergodic Markov Chain[7], and is guaranteed to converge to a stationary distribution regardless of where it begins [34]. Hence, the rank of each node is its value in the limit $\mathbf{w} := \lim_{t \to \infty} \mathbf{w}_t$ of the iteration process. Another nice consequence of this formulation is that the transition matrix given by $M := \lambda A^{\mathrm{T}} + (1 - \lambda)/N$ has eigenvalue 1, i. e., $M\mathbf{w} = 1\mathbf{w}$, therefore the right principal eigenvector of the matrix $M$, the one with eigenvalue 1, corresponds exactly to the fixed point of Equation 4.25. Among the existing methods for computing principal eigenvectors, the power iteration is the most typical one. Note that although the convergence is guaranteed, one can stop the algorithm earlier if the results are already satisfactory.

**PageRank for Folksonomies.** Since folksonomies can be represented as hypergraphs (see Section 1.4.2), PageRank cannot be applied directly. This is because the original PageRank algorithm is designed to be applied in graphs of edge size two.[8] Therefore, in order to apply the PageRank algorithm to folksonomies, Hotho et al.[17] proposed to convert the hypergraph $G_{\mathbb{F}} := (V, E)$ into an undirected tri-partite graph where the co-occurrences between tags and users, users and resources, and tags and resources, become edges between the respective nodes. I. e., each triple $(u, r, t)$ in $Y$ results in three undirected edges $\{u, r\}$, $\{u, t\}$, and $\{r, t\}$ in $E$ (see Figure 4.9).

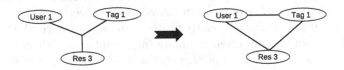

**Fig. 4.9** Converting a hyperedge into three undirected edges.

---

[6] A row-stochastic matrix refers to a matrix where the row-wise sum of elements, for each row in the matrix, is 1.

[7] A Markov chain is called an ergodic chain if it is possible to go from every state to every state.

[8] The size of an edge corresponds to the number of vertices it contains.

After the hypergraph transformation, one has the required setup to apply the original PageRank algorithm [2]. The following modifications must be performed though. First, the row-stochastic matrix $A \in \mathbb{R}^{N \times N}$, where $N :=$ $|U| + |R| + |T|$, is defined in terms of the co-frequency matrix $A'$, that is, $A'_{i,j} := c$ if $i$ and $j$ co-occur exactly $c$ times. For a global ranking, one will typically choose $\mathbf{p} = \mathbb{1}$, i.e., the vector containing $|A|$ many 1's. In order to generate recommendations, however, $\mathbf{p}$ can be tuned by giving a higher weight to the target entity for which one wants to generate a recommendation. The tag recommendation $\hat{T}_{u,r}$, for example, is then the set of top-$n$ nodes in the ranking, restricted to tags.

A remarkable difference between the application of PageRank in folksonomies and its typical application to web pages, is that the graph $G_{\mathbb{F}}$ is undirected. As a side effect, most of the weight that goes through an edge at moment $i$ will flow back at $i + 1$. The results are thus rather similar to a ranking that is simply based on frequency of vertices' co-occurrences.[9] This was experimentally observed by [17], who showed that the final ranking is biased by the global graph structure. As a consequence, [17] developed the following approach.

**FolkRank.** The undirectedness of graph $G_{\mathbb{F}}$ together with the typical skewed co-occurrence distribution between the entities of the folksonomy, makes it very difficult for the non-popular nodes to become highly ranked. This problem is solved by the FolkRank approach, which is described as follows:

1. Let $\mathbf{w}^{(0)}$ be the stationary probability distribution resulting from applying Equation 4.25 with $\mathbf{p} = 1$. We will call this method *GlobalRank*.
2. Let $\mathbf{w}^{(1)}$ be the stationary probability distribution resulting from applying Equation 4.25 with $\mathbf{p}[u] := 1 + |U|$, $\mathbf{p}[r] := 1 + |R|$, and $\mathbf{p}[v] := 1$ for $v \neq u, r$.
3. $\mathbf{w} := \mathbf{w}^{(1)} - \mathbf{w}^{(0)}$ is the final weight. The resulting weight $\mathbf{w}[k]$ of an element $k$ of the folksonomy is then called the *FolkRank* of $k$ [17].

**Multi-mode Recommendations** For generating tag recommendations for a given user/resource pair $(u, r)$, we compute the ranking as described and the tag nodes yield the recommendation scores

$$\hat{s}(u, r, t) := \mathbf{w}|_T$$

Similarly, one can compute recommendations for users (or resources) by giving preference to a certain user (or resource). Since FolkRank computes a ranking on all three dimensions of the folksonomy, this produces the most relevant tags, users, and resources for the given user (or resource).

**Remarks on Complexity** One iteration of the adapted PageRank requires the computation of $dA\mathbf{w}+(d-1)\mathbf{p}$, with $A \in \mathbb{R}^{s \times s}$ where $s := |U|+|R|+|T|$. If

---

[9] It is easy to show that the results are exactly equal to the frequency of vertices' co-occurrences in case of $\lambda = 1$.

$t$ marks the number of iterations, the complexity would therefore be $(s^2+s)t \in O(s^2t)$. However, since $A$ is sparse, it is more efficient to go linearly over all *tag assignments* in $Y$ to compute the product $A\mathbf{w}$. After rank computation we have to sort the weights of the tags to collect the top-$n$ tags.

### 4.2.2 Relational Neighbors for Tag Recommendations

Marinho et al. [35] proposed an approach based on a simple relational neighbor classifier for tag recommendations. The idea is to represent the set $P := \{(u,r) \mid \exists t \in T : (u,r,t) \in Y\}$ of all distinct user/resource combinations (*posts*) in $Y$ as a homogeneous,[10] undirected graph $G := (P,E)$. Edges can be annotated with a weight $w : P \times P \rightarrow \mathbb{R}$ representing the strength of the relation. It is assumed that vertices are related to each other if they share the same user:

$$\mathcal{R}_u := \{(p,p') \in P \times P \mid u(p) = u(p')\}$$

the same resource:

$$\mathcal{R}_r := \{(p,p') \in P \times P \mid r(p) = r(p')\}$$

or either share the same user or resource:

$$\mathcal{R}_u^r := \mathcal{R}_u \cup \mathcal{R}_r$$

Figure 4.10 depicts all these three different relational graphs. For convenience, let $u(p) := u$ and $r(p) := r$ denote the user and resource of post $p \in P$ respectively. Thus, each vertex is connected to each other either in terms of other users who tagged the same resource, or the resources tagged by the same user.

This graph representation enables to exploit the different relations between users in a more fundamental way. As a side effect, we gain tools from relational classification that can be directly applied to tag recommendations. Relational classification refers to an active area of machine learning where classifiers usually consider, additionally to the typical attribute-value data of objects, relational information. A scientific paper, for example, can be connected to another paper that has been written by the same author or because they share common citations. It has been shown in many classification problems that relational classifiers can perform better than purely attribute-based classifiers [6, 32, 38].

If we consider the graph $G := (P, \mathcal{R}_r)$ as input, we can easily derive the projection-based CF models described in Chapter 3. For that, we simply need

---

[10] By homogeneous we mean that there is only a binary relation $\mathcal{R} \subseteq P \times P$ between objects of the same type.

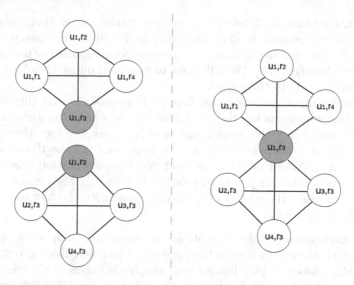

**Fig. 4.10** $\mathcal{R}_u$ (top left), $\mathcal{R}_r$ (bottom left) and $\mathcal{R}_u^r$ (right) of a given target node (colored node).

to weigh the edges of the graph with the cosine similarity between the user profile vectors (w. r. t. tags or resource) of neighboring nodes, i. e., $w(p, p') := \text{sim}\left(\mathbf{x}_{u(p)}, \mathbf{x}_{u(p')}\right)$ with $p' \in N_p$ with $N_p$ denoting the set of neighbors of node $p$. Then, for computing the set $\hat{T}_{u,r}$ of top-$n$ tags to recommend for a given target node $p \in P$, we would have:

$$\hat{T}_{u,r} := \underset{t \in T}{\arg\max} \sum_{\{p' \in N_p \mid t \in T(p')\}} w(p, p') \tag{4.26}$$

For the sake of simplicity, we will assume that the number of $k$ best neighbors of a specific $u(p)$ in the formula (4.26) above equals $|N_p|$.

We can easily derive a probabilistic version of CF by just normalizing each tag weight by the sum of similarities between the target node and its neighbors. So, the probability of a particular $t$ given a target node $p \in P$ is calculated by

$$P(t \mid p) = \frac{\sum_{p' \in N_p : t \in T(p')} w(p, p')}{\sum_{p' \in N_p} w(p, p')} \tag{4.27}$$

Now, for computing the top-$n$ recommendation list we just need to sort the tags in descending order of their probabilities, i. e.,

$$\hat{T}_{u,r} := \underset{t \in T}{\arg\max}^n P(t \mid p) \tag{4.28}$$

In fact, Equation 4.27 refers to a simple statistical relational classification method introduced in [33], that, despite its simplicity, has performed surprisingly well in many domains, even outperforming many other more sophisticated classifiers [33]. We will refer to this method as *WeightedAverage* (or WA form short) in the sequel.

A shortcoming of the projection-based CF methods is that they only exploit the tag assignments of the neighbors, but never the tag assignments of the target user. Since the users might want to reuse the tags they already used for other resources in the past, it is important to take those tags into account as well. Note that with the post graph representation this becomes straightforward, since we just need to apply WA on $\mathcal{R}_u^r$. The remaining question is how to weigh the edges in $\mathcal{R}_u$ since the similarity of a user with himself will always be maximal.

**Weight Estimation.** Edge weights are an important component of the relational graph since they denote the degree to which neighboring vertices are related. Marinho et al. [35] defined edge weights in terms of the cosine similarity between node profile vectors. In the following we define different node profile vectors that cover any situation into which two neighboring nodes can incur.

1. For two nodes $(p, p') \in \mathcal{R}_r$, we can represent $p$ and $p'$ as a binary user-tag profile vectors, where each component is 1, if a tag co-occurred with $u(p)$, or zero otherwise, i. e.,

$$\phi^{\text{user-tag}} := \left( \delta \left( u(p), R_{u(p)}, t \right) \right)_{t \in T}$$

Recall that $R_u$ corresponds to the set of all unique resources co-occurring with user $u$, and that $\delta(u, r, t) := 1$ if $(u, r, t) \in Y$, and 0 else. Since a user can use the same tag for several resources, we could alternatively consider the count of co-occurrences between users and tags as components of the vector, i. e.,

$$\phi^{\text{user-tag}} := (|Y \cap (\{u(p)\} \times R \times \{t\})|)_{t \in T}$$

2. If two users do not share any tags, we can still define their similarities in terms of user-resource profile vectors, i. e.,

$$\phi^{\text{user-res}} := \left( \delta \left( u(p), r, T_{u(p)} \right) \right)_{r \in R}$$

Note that it does not make much sense to define the user-resource vector components as frequency of co-occurrences, since a user only upload a resource once. Also note that these two cases correspond to the user profile vectors defined in terms of the projection matrices $\pi_{UT} Y$ and $\pi_{UR} Y$ respectively, introduced in Chapter 3 .

3. If $(p, p') \in \mathcal{R}_u$, we can not apply the vectors defined above since the cosine similarity between a user and himself will always be 1. Hence, we need to compute the weight in terms of binary resource-tag profile vectors

$$\phi^{\text{res-tag}} := \left( \delta \left( U_{r(p)}, r(p), t \right) \right)_{t \in T}$$

or co-frequency resource-tag profile vectors

$$\phi^{\text{res-tag}} := \left( |Y \cap (U \times \{r(p)\} \times \{t\})| \right)_{t \in T}$$

4. As well as for users, if two resources do not share any tags, we still can try to define their similarities in terms of binary resource-user profile vectors, i.e.,

$$\phi^{\text{res-user}} := \left( \delta \left( u, r(p), T_{r(p)} \right) \right)_{u \in U}$$

The edge weight is finally computed by applying the cosine similarity between the desired profile vectors:

$$\text{sim}(\phi(p), \phi(p')) := \frac{\langle \phi(p), \phi(p') \rangle}{\|\phi(p)\| \|\phi(p')\|} \tag{4.29}$$

**Remarks on Complexity.** Assuming that the weighted relational graph is given, the complexity of WA depends on two steps: (i) computation of tag weighted sums, which means $|N_p|$ passes in $T$, and (ii) computation of tag probabilities by normalizing the tag weighted sums, which means a pass in $T$. Hence, the whole complexity is given by:

$$O\left( |T|(|N_p| + \log(n) + 1) \right) \tag{4.30}$$

This cost can be reduced by dropping the normalization of tag weights, although at the cost of loosing a probabilistic interpretation.

## 4.3 Content and Social-Based Models

The recommendation task in a social tagging system can be improved by exploiting additional sources of information. This section presents an overview of methods that exploit the content of resources and the social relations between users as such additional sources.

### 4.3.1 Exploiting the Content of Resources

Relying solely on social tags for the task of locating relevant resources, can become problematic when the tags are ambiguous or overly personalized. An additional problem is the skewness in the occurrence frequency of tags, which in the extreme case results in individual tags used once per resource.

In a social tagging platform, the problem of automatic tag recommendation can be explored by considering the fact that user-provided tags can possibly reflect various aspects of the content of the resources that these tags are assigned to. This allows for developing tag-propagation schemes principled based on the intuition that 'similar' resources may contain relevant tags. Such an approach defines the *retrieval-based paradigm* [52] that first retrieves a set with the most similar resources to a query resource from the social tagging platform, and then assigns the query resource with a set of most relevant tags associated with the set of most similarly retrieved resources.

Content-based tag recommendation methods exploit a concept of similarity between resources according to their content, in order to apply a propagation of tags between similar resources. The *tagRelevance* method [30] follows such an approach, by proposing a neighbor-voting algorithm for tag relevance learning. The method propagates common tags through links introduced by content-based similarity between resources. Each tag accumulates its *relevance credit* by receiving neighbors' votes and, thus, tag relevance is estimated by counting neighbors' votes on it. Li et al. [30] show that under some realistic assumptions, *tagRelevance* produces a good tag relevance measurement for tag ranking. This result is also supported by experimental results on 3.5 million Flickr images for both social image retrieval and image tag suggestion.

Jeon et al. [19] propose an improved way of recommending tags based on similarity that is computed from the resources' content. Their method first finds the neighboring resources using a subject-based content analysis of the resources, as the subject of a resource commonly determines the tags provided by users about it. Next, this method applies a weighted neighbor voting technique. In contrast to the neighbor voting technique of Li et al. [30], which considers the tags of most votes among tags in similarity-based neighbor resources, the method of Jeon et al. [19] first filters the tags of low votes and extracts relevant tags by measuring the tag relevance using weights based on a search score that results from the content analysis. In the final step, the most relevant tags for a resource are predicted based on a tag-ranking scheme. Jeon et al. [19] develop their method for image data. To measure similarity they use the Euclidean distance between visual features that are extracted from the images by first identifying subject regions, using a graph-based visual saliency method. Next, the method finds tags with more than the threshold votes. Tag relevance is computed for each extracted tag by considering the similarity between resources. The relevance $\mathrm{rel}(t, r)$ between a tag $t$ and a resource $r$ is computed as:

$$\mathrm{rel}(t, r) = 1 - \frac{1}{|R_{N_{r,t}}|} \sum_{r' \in R_{N_{r,t}}} d(r, r'), \qquad (4.31)$$

where $N_r$ denotes the content-based $k$ nearest neighbors resources of resource $r$, $R_{N_{r,t}}$ denotes the set of resources in $N_r$ that have been assigned the tag $t$, and $d(r, r')$ the distance between the resources $r$ and $r'$ in the feature space constructed based on their content. Finally, considering each tag $t^*$ in the set of most relevant tags, which are computed with the score of Equation 4.31, tags are recommended to a resource $r$ by using the following tag-ranking score:

$$s(t^*) = \frac{1}{|R_{N_{r,t^*}}|} \sum_{r' \in R_{N_{r,t^*}}} \frac{o(t^*, T_{r'})}{|T_{r'}|}, \tag{4.32}$$

where $o(t^*, T_{r'})$ is the order of $t^*$ in $T_{r'}$.

Both the approaches by Li et al. [30] and Jeon et al. [19] create a feature-representation scheme to extract salient content features from resources, and then are using a standard distance measure to compute distances between resources within the created feature space. Differently from these approaches, Wu et al. [52] propose a unified *Distance Metric Learning* (DML) framework for the purpose of improving the accuracy of tag recommendations. Their DML approach learns metrics from implicit, side information hidden in resources. The reason for this is that, in contrast to conventional DML methods, side information in social tagging systems is not provided explicitly, neither in the form of pairwise constraints nor as class labels. In a social tagging system, side information is only implicitly available. As a result, the proposed unified DML method exploits both the information provided by the social tagging system and the content of the resources in order to develop an effective metric.

## 4.3.2 Exploiting Social Relations

Most recommender systems do not take into account the relations declared explicitly by users, such as, e. g., the "friendship" relations that some STS support. Therefore, these relations can be exploited to improve the task of recommendation. The intuition is that, related users (e. g., "friends") influence each other in terms of their tagging behavior.

Liu et al. [31] propose to exploit the social relations between users. In addition to that, following an approach that resembles the content-based approaches that are presented in Section 4.3.1, they inject content similarities between resources into the graph representation of social tagging data.

To measure the similarity among users, it is intuitive to assume that two users are more similar if they share several direct or indirect social connections (e. g., "friendships"). By following a standard representation of users' social relations as a feature vector (i. e., representing the social connections between users with a two-dimensional matrix), the similarity between two users is determined only based on their direct relations (e. g., "friend") and ignores intermediate relations to other users. To resolve this problem, Liu et al. [31] examine several methods to compute similarity based on random-walks over nodes in a graph. For two nodes $i$ and $j$ of a graph whose adjacency matrix is $A$, they consider the following similarity measures:

Average first-passage time (FPT): In this case, the similarity $\text{sim}(i,j) = m(i \mid j)$ is defined as the average number of steps for a random walker starting in $j$ to enter $i$. FPT is computed with $m(i \mid j) = 1 + \sum_{k=1}^{n} P_{j,k} m(i \mid k), j \neq i$ (and 0 for $j = i$).

Pseudoinverse of the Laplacian matrix $(L^+)$: The Laplacian matrix $L$ of the graph is defined as $L = D - A$ where $D_{i,i} = \sum_j A_{i,j}$ and $D_{i,j} = 0$ for $i \neq j$. Matrix $L^+$ is the Moore-Penrose pseudoinverse of $L$.

Matrix-forest-based algorithm (MFA): The similarity matrix in this case is defined as $T = (I + L)^{-1}$.

Using a random-walk based similarity measure, a personalized collaborative filtering can combine both the collaborative information and the personalized tag preferences. In particular, personalized-CF generates the top-$n$ recommended tags as follows:

$$\hat{T}_{u,r} = \underset{t \in T}{\arg\max}\, \lambda \cdot c(u,r,t) + (1 - \lambda) \cdot p(u,t), \qquad (4.33)$$

where $\lambda \in [0,1]$ is a user-defined parameter, $c(u,r,t)$ is the collaborative information that is personalized with the user's tag preference:

$$c(u,r,t) = \sum_{u' \in N_{u,r}} \text{sim}(u,u') \cdot \delta((u',r,t) \in Y) \cdot \text{sim}(u',t), \qquad (4.34)$$

having $p(u,t) = \text{sim}(u,t)$ and $N_{u,r}$ denotes the the top-$k$ similar users of $u$ who also have tagged $r$.

In a similar approach, Jiang et al. [20] propose a tag recommendation strategy based on *social comment* context. A social comment network describes the user, resources, tags, and the comments added to resources. Such a network forms an alternative way to represent users' social relations, since making comments on the information resources of other users is a popular activity on social networks nowadays. In such cases, the implicitly expressed (through comments) common interests allow for formulating implicit groups of users. In their method, Jiang et al. [20] represent the users with a directed, complete graph structure (each user corresponds to a node in this graph and there exists an edge between each pair of users), whereas each edge from node $i$ to node $j$ is weighted according to the number of comments (normalized by the total number of comments by all users) that user $i$ provided to content of user $j$. Based on this representation, it is proposed to quantify the influence of each user, i.e., the fact that the user's resources receive many in-links, with a measure of prestige $\text{Pres}(u_i)$ for each user $u_i$ that is defined as follows:

$$\text{Pres}(u_i) = \frac{\sum_{i=1}^{m-1} \sum_{j=1}^{n} C(u_i, j)}{n}, \qquad (4.35)$$

where $m$ is the total number of users, $n$ is the total number of resources contributed by user $u$, $j$ represents the resources, and $C(u_i, j)$ is the number of comments that user $u_i$ added to resource $j$. Based on the above measure, users are ranked according to their prestige.

For the recommendation of tags, tag co-occurrence between two tags $t_i$ and $t_j$ is measured by $P(t_j \mid t_i)$:

$$P(t_j \mid t_i) = \frac{|t_i \cap t_j|}{|t_i|}, \qquad (4.36)$$

where $|t_i \cap t_j|$ denotes the number of resources that have been assigned with both $t_i$ and $t_j$. This way, tags are selected for recommendation based on a boosting factor that weighs tag co-occurrence when the tag pairs have been used by a user with higher prestige.

Along this line, Rae et al. [39] propose a framework for combining information from various contexts, including the content of the resources, the social contacts of users, which include either direct connections (e. g., "friendship") to other individual users or membership to groups of users. Additional information from the social network is exploited in this framework for the task of tag recommendation as well as for expanding queries of users of social networks.

A significant difference of the approach proposed by Rae et al. [39] compared to the previously examined methods, is that it considers a series of layers in terms of the contextual information that is extracted from the social network; Figure 4.11 depicts the interactions between these contextual layers:

1. The first contextual layer is the *personal* (PC), which contains the tags of each individual user.
2. The second contextual layer is the *social contact context* (SCC), which is constructed by aggregating, for each specific user $u$, the tags of all users that are connected to $u$.
3. The third contextual layer is the *social group context* (SGC), which is constructed by aggregating, for each specific user $u$, the tags of resources belonging to (e. g., posted by) groups that $u$ is member of.
4. The fourth and final contextual layer is the *collective context* (CC), which is constructed by aggregating the tags of all users.

Within a probabilistic prediction framework proposed by Rae et al. [39], each contextual layer produces a weighted network of tags, having an edge between tags, when they have been used to annotate the same resource. For each tag $t$, its *occurrence tally* $o(t_i)$ is defined as the number of times that $t_i$ occurs (i. e., used to annotate resources). The weights of the edges in the network are determined by the co-occurrence tally $c(t_i, t_j)$, which is the number of times that tags $t_i$ and $t_j$ are used to annotate the same resource. Recommendations are produced based on a given set of (input) query tags. Each query tag generates first an intermediate set of recommendations that are

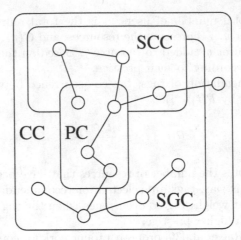

**Fig. 4.11** The interactions between the series of contextual layers (Figure adapted from [39]).

later combined. The set of recommendations for a given query tag and for a given context is then determined by the complete set of tags that co-occur with that tag in that context's network. The tags are ranked and penalization is applied to those tags that are not recommended by all query tags. In particular, given a set of query tags $Q$, the ranking score $p_x(t \mid Q)$ (probability) of a tag $t$ for an intermediate suggestion within a contextual layer $x$ is defined as:

$$p_x(t \mid Q) = p_x(t) \prod_{q \in Q} p_x(t \mid q), \tag{4.37}$$

where for $p(t \mid q)$ in general (i.e., for each context) it holds that $p(t \mid q) = c(t, q)/o(q)$. For the $t, q$ pairs for which $p(t \mid q) = 0$, we assign $p(t \mid q) = \varepsilon$, where $\varepsilon$ is a small positive quantity. According to the $p_x(t \mid Q)$ quantity, tags are ranked in descending order and the top-$n$ tags are recommended as given by that context's network of tags for a given query tag set. This approach can be used in the same way for each type network that can be constructed based on the aforementioned contextual layers. Experimental results in [39] indicate the advantage of combining different contextual layers.

In a recent investigation, Chidlovskii et al. [8] suggested a combination of content and social-based models for tag recommendation in social media systems. They propose an extension of the $k$-nearest neighbors ($k$-NN) classifier that can predict tags for a given resource, by observing the $k$ most similar neighbor resources. To avoid exhaustive similarity checks for all pairs of resources, they define the concept of *candidate set* that opts to a nearly optimal size-similarity trade-off that does not compromise the difference from truly similar resources. To compute candidate sets, Chidlovskii et al. propose to learn a local distance metric that takes into account both the personal space and social context. The personal space of a user is defined as the set of resources owned by her (in social media sites content is owned by users), whereas the social context of a user is defined by all resources owned by her contacts (e. g., 'friends'). Specifically, the proposed method applies the $k$-NN classifier for the case of multi-tag classification based on a Mahalanobis distance metric between the feature vectors describing the resources within the personal space and social context.

## 4.4 Further Reading

In addition to the methods presented in this chapter, there are many other interesting approaches incorporating other types of information sources as background knowledge into the recommendation models. Kim [24], for example, presents a method for personalized recommendation services using a tagging ontology for a social e-learning system. This method, based on reasoning rules, finds users with similar interests and clusters them in order to recommend resources. Cantador et al. [4] present a mechanism to automatically filter and classify raw tags in a set of purpose-oriented categories and find the underlying meanings (concepts) of the tags, by mapping them to semantic entities belonging to external knowledge bases.

Zheng and Li [54] investigate the importance and usefulness of tag and time information when predicting users' preference and examine how to exploit such information to build an effective resource-recommendation model. Kim et al. [22] propose leveraging user-generated tags as preference indicators and develop a new collaborative approach that first discovers relevant and irrelevant topics for users, and then enriches an individual user model with collaboration from other similar users. Zhou et al. [55] propose a tag-graph based community detection method to model the users' personal interests, which are further represented by discrete topic distributions.

Guy et al. [13] propose a new personalized resource recommendation method within an enterprise social media application suite that includes blogs, bookmarks, communities, wikis, and shared files. Hu et al. [18] propose to use distributional divergence to measure the similarity between users and examine two variations of such divergence (similarity) measures. Finally, in a recent survey paper, Gupta et al. [12] summarize different techniques employed to study various aspects of tagging, including recommendation methods.

An emerging research direction is the extraction and exploitation of semantics from social tags. Cantador et al. [3] proposed a method for the identification of semantic meanings and contexts that tags have within a particular folksonomy, in order to exploit them to build contextualized tag-based user and item profiles that allow item recommenders to achieve better precision and recall on their predictions. Kim et al. [23] proposed the use of tags for discovering relevant and irrelevant topics for users, in order to enrich models of individual users with collaboration from other similar users. Xu et al. [53] proposed a Semantic Enhancement Recommendation strategy (SemRec), based on both structural information and semantic information through a unified fusion model, in order to take into account important implicit semantic relationships hidden in tagging data. Finally, Li et al. [29] follow an approach that is motivated by the need to emulate human tagging behavior when recommending tags. In their approach they consider resources that are documents and take into account the semantics of the documents by discovering concepts contained in them. They represent each document using some few (most) relevant concepts derived from the concept space of Wikipedia. The recommendation of tags is based on the tag-concept model derived from the annotated resources of each tag.

# References

1. Michael W. Berry, Susan T. Dumais, and Gavin W. O'Brien. Using linear algebra for intelligent information retrieval. *SIAM Rev.*, 37:573–595, December 1995.
2. Sergey Brin and Lawrence Page. The anatomy of a large-Scale hypertextual web search engine. *Computer Networks and ISDN Systems*, 30(1-7):107–117, April 1998.
3. Iván Cantador, Alejandro Bellogín, Ignacio Fernández-Tobás, and Sergio López-Hernández. Semantic contextualisation of social tag-based profiles and item recommendations. In *E-Commerce and Web Technologies*, pages 101–113. 2011.
4. Iván Cantador, Ioannis Konstas, and Joemon M. Jose. Categorising social tags to improve folksonomy-based recommendations. *Web Semant.*, 9:1–15, 2011.
5. J.D. Carroll and J.J. Chang. Analysis of individual differences in multidimensional scaling via an n-way generalization of eckart-young decomposition. *Psychometrika*, 35:283–319, 1970.
6. Soumen Chakrabarti, Byron Dom, and Piotr Indyk. Enhanced hypertext categorization using hyperlinks. *SIGMOD Rec.*, 27(2):307–318, 1998.
7. Shouchun Chen, Fei Wang, and Changshui Zhang. Simultaneous heterogeneous data clustering based on higher order relationships. In *Proceedings of the Seventh IEEE International Conference on Data Mining Workshops*, ICDMW '07, pages 387–392, Washington, DC, USA, 2007. IEEE Computer Society.
8. Boris Chidlovskii and Aymen Benzarti. Local metric learning for tag recommendation in social networks. In *Proceedings of the 11th ACM symposium on Document engineering*, DocEng '11, pages 205–208, 2011.

9. Scott Deerwester, Susan T. Dumais, George W. Furnas, Thomas K. Landauer, and Richard Harshman. Indexing by latent semantic analysis. *Journal of the American Society for Information Science*, 41(6):391–407, 1990.

10. P. Drineas and M. W. Mahoney. A randomized algorithm for a tensor-based generalization of the svd. *Linear Algebra and Its Applications*, 420(2–3):553–571, 2007.

11. G. W. Furnas, S. Deerwester, S. T. Dumais, T. K. Landauer, R. A. Harshman, L. A. Streeter, and K. E. Lochbaum. Information retrieval using a singular value decomposition model of latent semantic structure. In *Proceedings of the 11th annual international ACM SIGIR conference on Research and development in information retrieval*, SIGIR '88, pages 465–480, New York, NY, USA, 1988. ACM.

12. Manish Gupta, Rui Li, Zhijun Yin, and Jiawei Han. Survey on social tagging techniques. *SIGKDD Explor. Newsl.*, 12:58–72, November 2010.

13. Ido Guy, Naama Zwerdling, Inbal Ronen, David Carmel, and Erel Uziel. Social media recommendation based on people and tags. In *Proceeding of the 33rd international ACM SIGIR conference on Research and development in information retrieval*, SIGIR '10, pages 194–201, 2010.

14. R.A. Harshman. Foundations of the parafac procedure: models and conditions for an 'exploratory' multimodal factor analysis. In *UCLA Working Papers in Phonetics*, pages 1–84, 1970.

15. Thomas Hofmann. Latent semantic models for collaborative filtering. *ACM Transactions on Information Systems*, 22(1):89–115, January 2004.

16. Andreas Hotho, Dominik Benz, Robert Jäschke, and Beate Krause, editors. *ECML PKDD Discovery Challenge 2008 (RSDC'08)*. Workshop at 18th Europ. Conf. on Machine Learning (ECML'08) / 11th Europ. Conf. on Principles and Practice of Knowledge Discovery in Databases (PKDD'08), 2008.

17. Andreas Hotho, Robert Jäschke, Christoph Schmitz, and Gerd Stumme. Information retrieval in folksonomies: Search and ranking. In York Sure and John Domingue, editors, *The Semantic Web: Research and Applications*, volume 4011 of *Lecture Notes in Computer Science*, pages 411–426, Berlin/Heidelberg, June 2006. Springer.

18. Meiqun Hu, Ee-Peng Lim, and Jing Jiang. A probabilistic approach to personalized tag recommendation. *Social Computing / IEEE International Conference on Privacy, Security, Risk and Trust, 2010 IEEE International Conference on*, pages 33–40, 2010.

19. Won Jeon, Sunyoung Cho, Jaeseong Cha, and Hyeran Byun. Tag suggestion using visual content and social tag. In *Proceedings of the 5th International Conference on Ubiquitous Information Management and Communication*, ICUIMC '11, pages 104:1–104:5, 2011.

20. Bo Jiang, Yun Ling, and Jiale Wang. Tag recommendation based on social comment network. *Journal of Digital Content Technology and its Applications*, 4(8):110–117, 2010.

21. Robert Jäschke, Leandro Balby Marinho, Andreas Hotho, Lars Schmidt-Thieme, and Gerd Stumme. Tag recommendations in folksonomies. In Joost N. Kok, Jacek Koronacki, Ramon López de Mántaras, Stan Matwin, Dunja Mladenic, and Andrzej Skowron, editors, *Knowledge Discovery in Databases: PKDD 2007, 11th European Conference on Principles and Practice of Knowledge Discovery in Databases*, volume 4702 of *Lecture Notes in Computer Science*, pages 506–514, Berlin, Heidelberg, 2007. Springer.

22. Heung-Nam Kim, Abdulmajeed Alkhaldi, Abdulmotaleb El Saddik, and Geun-Sik Jo. Collaborative user modeling with user-generated tags for social recommender systems. *Expert Syst. Appl.*, 38:8488–8496, 2011.

23. Heung-Nam Kim, Abdulmajeed Alkhaldi, Abdulmotaleb El Saddik, and Geun-Sik Jo. Collaborative user modeling with user-generated tags for social recommender systems. *Expert Systems with Applications*, 38(7):8488 – 8496, 2011.

24. Hyon Hee Kim. A personalized recommendation method using a tagging ontology for a social e-learning system. In *Proceedings of the Third international conference on Intelligent information and database systems - Volume Part I*, ACIIDS'11, pages 357–366, 2011.

25. Tamara G. Kolda and Jimeng Sun. Scalable tensor decompositions for multi-aspect data mining. In *ICDM '08: Proceedings of the 8th IEEE International Conference on Data Mining*, pages 363–372. IEEE Computer Society, December 2008.

26. Yehuda Koren. Factorization meets the neighborhood: a multifaceted collaborative filtering model. In *KDD '08: Proceeding of the 14th ACM SIGKDD International Conference on Knowledge Discovery and Data Mining*, pages 426–434. ACM, 2008.

27. Yehuda Koren. Collaborative filtering with temporal dynamics. In *KDD '09: Proceedings of the 15th ACM SIGKDD International Conference on Knowledge Discovery and Data Mining*, pages 447–456. ACM, 2009.

28. Lieven De Lathauwer, Bart De Moor, and Joos Vandewalle. A multilinear singular value decomposition. *SIAM Journal on Matrix Analysis and Applications*, 21(4):1253–1278, 2000.

29. Chenliang Li, Anwitaman Datta, and Aixin Sun. Semantic tag recommendation using concept model. In *Proceedings of the 34th international ACM SIGIR conference on Research and development in Information*, SIGIR '11, pages 1159–1160, 2011.

30. Xirong Li, Cees G. M. Snoek, and Marcel Worring. Learning social tag relevance by neighbor voting. *IEEE Transactions on Multimedia*, 11(7):1310–1322, 2009.

31. Kaipeng Liu, Binxing Fang, and Weizhe Zhang. Speak the same language with your friends: augmenting tag recommenders with social relations. In *Proceedings of the 21st ACM conference on Hypertext and hypermedia*, HT '10, pages 45–50, 2010.

32. Qing Lu and Lise Getoor. Link-based classification using labeled and unlabeled data. In *Proceedings of the ICML Workshop on The Continuum from Labeled to Unlabeled Data in Machine Learning and Data Mining*, 2003.

33. Sofus A. Macskassy and Foster Provost. A simple relational classifier. In *MRDM '03: Proceedings of the Second Workshop on Multi-Relational Data Mining at KDD-2003*, pages 64–76, 2003.

34. Hinrich Schütze Christopher D. Manning. *Introduction to Information Retrieval.* ISBN 978-0-521-86571-5. Cambridge University Press, May 2008.

35. Leandro Balby Marinho, Christine Preisach, and Lars Schmidt-Thieme. Relational classification for personalized tag recommendation. In Folke Eisterlehner, Andreas Hotho, and Robert Jäschke, editors, *ECML PKDD Discovery Challenge 2009 (DC09)*, volume 497 of *CEUR-WS.org*, pages 7–16, 2009.

36. Alexandros Nanopoulos, Dimitrios Rafailidis, Panagiotis Symeonidis, and Yannis Manolopoulos. Musicbox: Personalized music recommendation based on cubic analysis of social tags. *IEEE Transactions on Audio, Speech and Language Processing*, 18(2):407–412, 2010.

37. Alexandros Nanopoulos Panagiotis Symeonidis, Maria Ruxanda and Yannis Manolopoulos. Ternary semantic analysis of social tags for personalized music recommendation. In *ISMIR '08: Proceedings of the 9th ISMIR Conference*, pages 219–224, New York, 2008.

38. Christine Preisach and Lars Schmidt-Thieme. Relational ensemble classification. In *ICDM '06: Proceedings of the 6th International Conference on Data Mining*, pages 499–509. IEEE Computer Society, 2006.

39. Adam Rae, Börkur Sigurbjörnsson, and Roelof van Zwol. Improving tag recommendation using social networks. In *Adaptivity, Personalization and Fusion of Heterogeneous Information*, RIAO '10, pages 92–99, 2010.

40. Steffen Rendle. *Context-Aware Ranking with Factorization Models.* Springer Berlin Heidelberg, 1st edition, November 2010.

41. Steffen Rendle. Factorization machines. In *Proceedings of the 10th IEEE International Conference on Data Mining.* IEEE Computer Society, 2010.

42. Steffen Rendle, Christoph Freudenthaler, Zeno Gantner, and Lars Schmidt-Thieme. BPR: Bayesian personalized ranking from implicit feedback. In *UAI '09: Proceedings of the 25th Conference on Uncertainty in Artificial Intelligence*, 2009.

43. Steffen Rendle, Leandro B. Marinho, Alexandros Nanopoulos, and Lars Schimdt-Thieme. Learning optimal ranking with tensor factorization for tag recommendation. In *KDD '09: Proceedings of the 15th ACM SIGKDD International Conference on Knowledge Discovery and Data Mining*, pages 727–736. ACM, 2009.

44. Steffen Rendle and Lars Schmidt-Thieme. Pairwise interaction tensor factorization for personalized tag recommendation. In *WSDM '10: Pro-*

*ceedings of the Third ACM International Conference on Web Search and Data Mining.* ACM, 2010.

45. Ruslan Salakhutdinov and Andriy Mnih. Bayesian probabilistic matrix factorization using markov chain monte carlo. In *ICML '08: Proceedings of the 25th International Conference on Machine Learning*, pages 880–887. ACM, 2008.

46. Amnon Shashua and Tamir Hazan. Non-negative tensor factorization with applications to statistics and computer vision. In *ICML '05: Proceedings of the 22nd International Conference on Machine Learning*, pages 792–799. ACM, 2005.

47. Jian-Tao Sun, Hua-Jun Zeng, Huan Liu, Yuchang Lu, and Zheng Chen. Cubesvd: a novel approach to personalized web search. In *Proceedings of the 14th international conference on World Wide Web*, WWW '05, pages 382–390, New York, NY, USA, 2005. ACM.

48. Panagiotis Symeonidis, Alexandros Nanopoulos, and Yannis Manolopoulos. A unified framework for providing recommendations in social tagging systems based on ternary semantic analysis. *IEEE Transactions on Knowledge and Data Engineering*, 22(2), 2010.

49. L. Tucker. Some mathematical notes on three-mode factor analysis. *Psychometrika*, pages 279–311, 1966.

50. P. Turney. Empirical evaluation of four tensor decomposition algorithms. *Technical Report (NRC/ERB-1152)*, 2007.

51. Hongcheng Wang and Narendra Ahuja. A tensor approximation approach to dimensionality reduction. *Int. J. Comput. Vision*, 76:217–229, March 2008.

52. Pengcheng Wu, Steven Chu-Hong Hoi, Peilin Zhao, and Ying He. Mining social images with distance metric learning for automated image tagging. In *Proceedings of the fourth ACM international conference on Web search and data mining*, WSDM '11, pages 197–206, 2011.

53. Guandong Xu, Yanhui Gu, Peter Dolog, Yanchun Zhang, and Masaru Kitsuregawa. Semrec: A semantic enhancement framework for tag based recommendation. In *Proceedings of the Twenty-Fifth AAAI Conference on Artificial Intelligence (AAAI)*, 2011.

54. Nan Zheng and Qiudan Li. A recommender system based on tag and time information for social tagging systems. *Expert Syst. Appl.*, 38:4575–4587, 2011.

55. Tom Chao Zhou, Hao Ma, Michael R. Lyu, and Irwin King. Userrec: A user recommendation framework in social tagging systems. In *Proceedings of the Twenty-Fourth AAAI Conference on Artificial Intelligence (AAAI)*, 2010.

# Chapter 5
# Offline Evaluation

In this chapter we present the most usual experimental protocols and metrics employed for offline evaluation of tag recommender systems. By offline we mean that the algorithms are evaluated on a snapshot of some real-world STS dataset, which, in turn, is typically split into training and test datasets. This corresponds to the most typical evaluation scenario found in the literature since researchers do not need to have a STS up and running for assessing the performance of his/her algorithms. We also summarize the main tag recommendation algorithms presented in this book, pointing out pros and cons in terms of the metrics and protocols introduced in this chapter.

## 5.1 Evaluation Metrics

In this section we present the typical metrics for evaluating the performance of tag recommendation algorithms. Notice that for an evaluation of tag-aware recommender systems one would employ the same evaluation protocols and metrics of traditional recommender systems, and therefore we do not present these here. For good surveys on evaluation of traditional recommender systems we refer to [2, 12].

### 5.1.1 Precision and Recall

Independent of the choice of 'gold standard' that represents the perfect tags a recommender should suggest for a certain post, there exist two very common measures that measure *what proportion of the recommended tags was correct* and *what proportion of the correct tags could be recommended*. These two measures – precision and recall, respectively – are standard in such scenarios [2]. For each post $(u, r, T_{u,r})$ from the test data, one compares the set

$\hat{T}_{u,r}$ of recommended tags with the set $T_{u,r}$ of true tags. Then, precision and recall of a recommendation are defined as follows:

$$\text{recall}(T_{u,r}, \hat{T}_{u,r}) = \frac{|T_{u,r} \cap \hat{T}_{u,r}|}{|T_{u,r}|}, \quad \text{precision}(T_{u,r}, \hat{T}_{u,r}) = \frac{|T_{u,r} \cap \hat{T}_{u,r}|}{|\hat{T}_{u,r}|} .$$

The 'true tags' are typically the tags the user assigned to the post, but could also be manually chosen tags, e.g., by an expert. For an empty recommendation $\hat{T}_{u,r} = \emptyset$, one typically sets precision$(T_{u,r}, \emptyset) = 0$.

Since for a fixed number of recommended tags precision and recall are dependent, one often restrains the evaluation to one of the measures, for example recall@5, i.e., the recall for five recommended tags. This accommodates the fact that social tagging systems often recommend a limited and in particular fixed number of tags. Another option is the use of the F1 measure (f1m), the harmonic mean of precision and recall: f1m = 2(precision · recall)/(precision + recall).

### 5.1.2 Further Measures

In contrast to the kind of 'gold standard evaluation' presented in the previous section stands the manual judgement of recommendations. It has the benefit of detecting and honoring correct or good recommended tags that were not chosen by the user or some expert to tag the resource. However, manual judgement is relatively expensive and time consuming and often only possible for a small number of recommendations.

For real world recommender systems other factors besides the quality of the recommendation are crucial, e.g., recommendation latency and algorithm complexity. The recommender system must be able to produce good recommendations in an acceptable amount of time, given the restricted memory and processing resources. Such issues can be addressed by an online evaluation, as shown in Section 7.3.

## 5.2 Evaluation Protocols

### 5.2.1 LeavePostOut Methodology

A variant of the leave-one-out hold-out estimation [2] called *LeavePostOut* is quite popular when evaluating tag recommendations against an offline gold standard dataset. To this end, one randomly picks, for each user $u$, a resource from $r_u$, which the user had posted before. The task of the recommender is then to predict the tags the user assigned to $r_u$, based on the gold standard dataset without the post $(u, r_u, T_{u,r_u})$. This process is then repeated, each time with a randomly chosen resource per user, to further minimize the variance. Recall and precision values are then averaged over all the runs.

### 5.2.2 Time-based Splits

Another option to hold out posts from the gold standard for testing the recommendation quality is to split the dataset by time. To this end, one choses a point in time that lies within the timespan of the dataset and then uses all posts before that point as training data and all posts after that point as test data. This more closely resembles the situation of a recommender in a real social tagging system, where it 'knows' only the previous posts at recommendation time. A variant of this method that almost exactly reproduces this behaviour is to start the evaluation of the recommender on the test data with the first post after the split point and then let the recommender use this post to improve its model. Then one evaluates the next post and so on. The case study in Section 7.3 as well as the online task of the Discovery Challenge both used this kind of evaluation since they measure the performance of the recommenders within a real social tagging system.

## 5.3 Comparison of Tag Recommenders

Many of the most important tag recommendation methods proposed so far were evaluated through the metrics and protocols presented in this chapter, which enables to compare them under a common basis, although not always the same dataset or train/test dataset splits are used across the original papers. In this section we summarize the tag recommendation algorithms presented so far pointing out pros and cons in terms of the evaluation protocols and metrics presented in this chapter. Note that we just consider the algorithms without attributes.

We saw in Section 3.3 that in order to apply standard neighborhood-based CF algorithms to folksonomies, some data transformation must be performed. Although these methods usually attain better recommendations than the baselines (cf. Section 3) [7], such transformations lead to information loss, i.e., one mode is always discarded, which can lower the recommendation quality. Another well known problem with CF-based methods is that large projection matrices must be kept in memory, which can make these algorithms unfeasible for large scale datasets. Furthermore, for each different mode to be recommended, the algorithm must be eventually changed, demanding an additional effort for offering multi-mode recommendations.

FolkRank also breaks the ternary relations of the folksonomy into binary relations in order to apply the PageRank algorithm (cf. Section 4.2.1). FolkRank proved to give significantly better tag recommendations than the baselines and CF in three different STS datasets [3, 4]. This method also enables easy switch of the recommendation mode, i. e., for recommending users (or resources) one can give preference to a certain user (or resource) and restrict the results to the set of top-$n$ user (or resource) nodes. Moreover, as well as CF-based algorithms, FolkRank is robust against online updates since it does not need to be trained every time a new user, resource or tag enters the system. However, FolkRank is computationally expensive, making it more suitable for systems where real-time recommendation is not a requirement.

Similarly to CF and FolkRank the WeightedAverage algorithm (cf. Equation 4.27) operates on binary relations between entities of the folksonomy. Since CF is a special case of this algorithm (cf. Sec. 4.2.2), it suffers from the same scalability issues as CF, i. e., it can be unfeasible to keep a relational graph in memory for large scale STS datasets. But apart from this restriction and once the graph is available, recommendations are fast to compute. Moreover, this algorithm proved to be competitive to the state-of-the-art achieving the second place in the offline tag recommendation task of the ECML PKDD Discovery Challenge 2009 [1].[1]

Tensor factorization methods work directly on the ternary relations of folksonomies, thus using the full information available. Although the learning phase can be costly, it can be performed offline. After the model is learned, the recommendations are fast to compute, making these algorithms suitable for real-time recommendations. HOSVD was one of the first tensor factorization approaches for tag recommendation. RTF (cf. Section 4.1.3) appeared next as a more specialized alternative to HOSVD, where the parameter learning is cast as a personalized tag ranking problem. Although having the same overall cost as HOSVD for the model training phase, RTF proved to have superior predictive power than HOSVD and the other aforementioned tag recommendation algorithms [9]. PITF was built on top of RTF, improving its training runtime while keeping similar predictive power. PITF turned out to be the winning approach of the offline tag recommendation task of the ECML PKDD Discovery Challenge 2009.

Factorization Machines (cf. Sec. 4.1.2.4) appear as a generalization of the most specialized state-of-the-art methods, such as RTF and PITF. They are generic in the sense that they can be applied to different recommendation tasks (e. g., resource and user) without requiring significant changes in the learning algorithm. Moreover, they feature competitive predictive power to PITF while keeping modest computational costs.

Figure 5.1 shows a comparison between some of the aforementioned algorithms in snapshots of BibSonomy and Last.fm [11]. Table 5.1 shows the

---

[1] Actually the submission to the challenge consisted of an ensemble between two similar methods [6]: *WeightedAverage* and *Probabilistic Weighted Average*, an iterative relational classifier introduced in [8].

official results for the top-3 methods of the offline tag recommendation task of the ECML PKDD Discovery Challenge 2009.

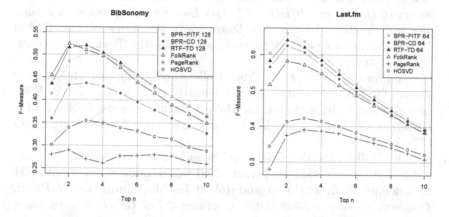

**Fig. 5.1** The tensor factorization models (RTF-TD, BPR-CD, BPR-PITF) compared against FolkRank, PageRank and HOSVD on snapshots of Last.fm and BibSonomy [11]. The legend also indicates the number of latent factors used for RTF and PITF in each dataset.

**Table 5.1** Official results of the ECML PKDD Discovery Challenge 2009.

| Rank | Method | Top-5 F1 |
|------|-----------------------|----------|
| 1 | PITF [10] | 0.35594 |
| 2 | Relational Ensemble [6] | 0.33185 |
| 3 | Content-based [5] | 0.32461 |

# References

1. Folke Eisterlehner, Andreas Hotho, and Robert Jäschke, editors. *ECML PKDD Discovery Challenge 2009 (DC09)*, volume 497 of *CEUR-WS.org*, September 2009.
2. Jonathan L. Herlocker, Joseph A. Konstan, Loren G. Terveen, and John T. Riedl. Evaluating collaborative filtering recommender systems. *ACM Trans. Inf. Syst.*, 22(1):5–53, 2004.
3. Robert Jäschke, Leandro Marinho, Andreas Hotho, Lars Schmidt-Thieme, and Gerd Stumme. Tag recommendations in social bookmarking systems. *AI Communications*, 21(4):231–247, 2008.

4. Robert Jäschke, Leandro Balby Marinho, Andreas Hotho, Lars Schmidt-Thieme, and Gerd Stumme. Tag recommendations in folksonomies. In Joost N. Kok, Jacek Koronacki, Ramon López de Mántaras, Stan Matwin, Dunja Mladenic, and Andrzej Skowron, editors, *Knowledge Discovery in Databases: PKDD 2007, 11th European Conference on Principles and Practice of Knowledge Discovery in Databases*, volume 4702 of *Lecture Notes in Computer Science*, pages 506–514, Berlin, Heidelberg, 2007. Springer.

5. Marek Lipczak, Yeming Hu, Yael Kollet, and Evangelos Milios. Tag sources for recommendation in collaborative tagging systems. In Folke Eisterlehner, Andreas Hotho, and Robert Jäschke, editors, *ECML PKDD Discovery Challenge 2009 (DC09)*, volume 497 of *CEUR-WS.org*, pages 157–172, 2009.

6. Leandro Balby Marinho, Christine Preisach, and Lars Schmidt-Thieme. Relational classification for personalized tag recommendation. In Folke Eisterlehner, Andreas Hotho, and Robert Jäschke, editors, *ECML PKDD Discovery Challenge 2009 (DC09)*, volume 497 of *CEUR-WS.org*, pages 7–16, 2009.

7. Leandro Balby Marinho and Lars Schmidt-Thieme. Collaborative tag recommendations. In *GFKL '07: Proceedings of the 31st Annual Conference of the Gesellschaft für Klassifikation (GfKl), Freiburg*, pages 533–540. Springer, 2007.

8. Christine Preisach, Leandro Balby Marinho, and Lars Schmidt-Thieme. Semi-supervised tag recommendation - using untagged resources to mitigate cold-start problems. In *PAKDD '10: Proceedings of the 14th Pacific-Asia Conference on Advances in Knowledge Discovery and Data Mining*, 2010. to appear.

9. Steffen Rendle, Leandro B. Marinho, Alexandros Nanopoulos, and Lars Schimdt-Thieme. Learning optimal ranking with tensor factorization for tag recommendation. In *KDD '09: Proceedings of the 15th ACM SIGKDD International Conference on Knowledge Discovery and Data Mining*, pages 727–736. ACM, 2009.

10. Steffen Rendle and Lars Schmidt-Thieme. Factor models for tag recommendation in BibSonomy. volume 497 of *CEUR-WS.org*, pages 235–242, 2009.

11. Steffen Rendle and Lars Schmidt-Thieme. Pairwise interaction tensor factorization for personalized tag recommendation. In *WSDM '10: Proceedings of the Third ACM International Conference on Web Search and Data Mining*. ACM, 2010.

12. Guy Shani and Asela Gunawardana. Evaluating recommendation systems. In Francesco Ricci, Lior Rokach, Bracha Shapira, and Paul B. Kantor, editors, *Recommender Systems Handbook*, pages 257–297. Springer US, 2011.

# Part III
# Implementing Recommender Systems for Social Tagging

# Chapter 6
# Real World Social Tagging Recommender Systems

As an exemplary implementation of a recommender system for social tagging systems we present in this chapter the tag recommendation framework of BibSonomy. It allows to test, evaluate and compare different tag recommendation algorithms in an online setting, where the users of BibSonomy actually see the recommendations during the posting process. The chapter is based on work published in [10, 11].

## 6.1 Introduction

The social tagging system BibSonomy[1] allows users to share bookmarks and publication references. It is developed and run by the Knowledge and Data Engineering Group Kassel where three of the authors of this book work(ed). BibSonomy was chosen as example to present a real world social tagging recommender system since it uses a well documented tag recommendation framework that was also used for evaluating tag recommendation approaches during the ECML PKDD Discovery Challenge 2009.

The design of the framework was motivated by the authors' research on tag recommendation algorithms and experiences from organizing the ECML PKDD Discovery Challenge 2008:

- The arguments to provide the user with tag recommendations discussed in Section 1.5 support the need for good recommendations in BibSonomy. Therefore, we needed a foundation to implement and run appropriate methods in the online system.
- The experience we gained by organizing the ECML PKDD Discovery Challenge 2008 showed that evaluation and comparison of different recommender systems in an offline setting can suffer from artifacts present in

---

[1] http://www.bibsonomy.org/

83

the data like masses of imported or automatically annotated posts. Furthermore, a realistic setting should force the recommenders to adhere to timeouts and other constraints which are difficult to control in an offline setting. Therefore, we needed a framework which allowed us the evaluation of online tag recommendations as one task of the Discovery Challenge 2009 we also organized (more on this in Section 7.3).

- We want to offer the tag recommendation research community a realistic testbed for their methods.
- Existing frameworks (cf. Section 6.6) mostly do not fit the tag recommendation scenario (Section 1.5.3) we have to handle (e. g., they do not suggest re-occurring items).

The framework is responsible for delivering tag recommendations to the user in two situations: when he edits a bookmark or publication post. Since the part of the user interface showing recommendations is very similar for both the bookmark posting and the publication posting page, we show in Figure 6.1 the relevant part of the 'postBookmark' page only.[2]

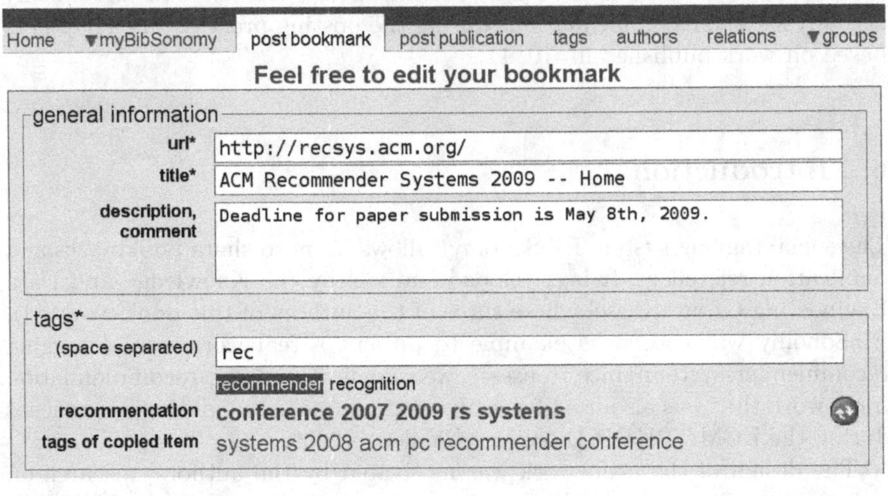

**Fig. 6.1** BibSonomy's recommendation interface on the bookmark posting page. The box labeled 'tags' contains a text input field where the user can enter the (space separated) tags, tags suggested for autocompletion, the tags from the recommender (bold), and the tags from the post the user just copies.

Below the fields for entering URL, title, and a description (which are typically automatically filled), the box labeled 'tags' keeps together the tagging information. There, the user can manually enter the tags to describe the resource. During typing she is assisted by a JavaScript autocompletion which

---

[2] Logged in users can access this page at http://www.bibsonomy.org/postBookmark.

selects tags among the recommended tags and all of her previously used tags whose prefix matches the already entered letters. The suggested tags are shown directly below the tag input box (in the screenshot *recommender*, *recognition*, and *recht*). Further down there are in bold letters up to five recommended tags ordered by their score from left to right. Thus, the recommender in action regarded *conference* to be the most appropriate tag for this resource and user. To the very right of the recommendation is a small icon depicting the *reload* button. It allows the user to request a new tag recommendation if she is unsatisfied with the one shown or wants to request further tags. We investigate the usage of this button in Section 7.2.2.

Besides triggering autocompletion with the tabulator key during typing, users can also click on tags with their mouse. They are then added to the input box. When the user copies a resource from another user's post, the tags the other user used to annotate the resource are shown below the recommended tags ('tags of copied item'). They are also regarded for autocompletion.

Aside from describing the framework we also try to answer such questions like "What is the performance of a recommender?", "Are there users with a tendency to a certain recommender?", or "Which click behaviour do users show?".

In this chapter we use the formalization of the tag recommendation task as introduced in Section 2.6 with a simplified notation for the scoring function $\hat{s}$, omitting the two variables $u$ and $r$.[3] The number of recommended tags is fixed to five throughout our analysis, although a recommender is allowed to return less than five tags.

This chapter is structured as follows: First, we present challenges and requirements to the framework in Section 6.2. Then, in Section 6.3, we briefly introduce the social tagging system BibSonomy. The description of the framework's architecture in Section 6.4 is followed by a presentation of available recommender implementations in Section 6.5. We conclude the chapter with a review of related work in the field in Section 6.6.

## 6.2 Challenges and Requirements

**Challenges.** For a recommender system to be successful in a real world application, it must approach several challenges. First, the provided recommendations must match the situation, i. e., tags should describe the annotated resource, products should awake the interest of the user, suggested resources should be interesting and relevant. Second, they must be delivered timely without delay and they must be easy to access (i. e., by allowing the user to click on them or to use tab-completion when entering tags). Third, the

---

[3] Although, of course, $\hat{s}$ also depends on $u$ and $r$, we will omit those two variables to simplify notation. Since $\hat{s}$ always appears together with the set of recommended tags $\hat{T}_{u,r}$, it should be clear from context, which $\hat{s}$ is meant.

system must ensure that recommendations do not impede the normal usage of the system.

**Technological and Infrastructure Requirements.** Implementing a recommendation service for a STS requires to tackle several problems, some of them we describe here.

First, having enough data available for recommendation algorithms to produce helpful recommendations is an important requirement one must address already in the design phase. The recommender needs access to the system's database and information about what the user is currently doing (which could be accomplished, e. g., by (re)-loading recommendations using techniques like AJAX). Further data – like the full text of documents – could be supplied to tackle the cold-start problem (e. g., for content-based recommenders). The system must be able to handle large amounts of data, to quickly select relevant subsets and provide methods for preprocessing.

The available hardware and expected amount of data limits the choice of recommendation algorithms which can be used. Although some methods allow (partial) precomputation of recommendations, this needs extra memory and might not yield the same good results as online computation, since recently added posts are missing in the model. Both hardware and network infrastructure must ensure short response times to deliver the recommendations to the user without too much delay. Together with a simple and non-intrusive user interface this ensures usability.

Further aspects which should be taken into account include implementation of logging of user events (e. g., clicking, key presses, etc.) to allow for efficient evaluation of the used recommendation methods in an online setting. Together with an online evaluation this also allows to tune the result selection strategies to dynamically choose the (currently) best recommendation algorithm for the user at hand.

## 6.3 The BibSonomy Social Tagging System

BibSonomy [1] started as a students project at the Knowledge and Data Engineering Group of the University of Kassel[4] in spring 2005. The goal was to implement a system for organizing BIBTEX [16] entries in a way similar to bookmarks in Delicious – which was at that time becoming more and more popular. BIBTEX is a popular literature management system for LATEX [13], which many researchers use for writing scientific papers. After integrating bookmarks as a second type of resource into the system and upon the progress made, BibSonomy was opened for public access at the end of 2005 – first announced to collegues only, later in 2006 to the public.

---

[4] http://www.kde.cs.uni-kassel.de/

A detailed view of one bookmark post in BibSonomy can be seen in Figure 6.2. The first line shows in bold the title of the bookmark which has the URL of the bookmark as underlying hyperlink. The second line shows an optional description the user can assign to every post. The last two lines belong together and show detailed information: first, all the tags the user has assigned to this post (*web, service, tutorial, guidelines* and *api*), second, the user name of that user (*hotho*) followed by a note, how many users tagged that specific resource. These parts have underlying hyperlinks, leading to the corresponding tag pages of the user, the users page and a page showing all four posts (i. e., the one of user *hotho* and those of the three other people) of this resource. The structure of a publication post is very similar, as seen in Figure 6.3.

**REST web services**
Good intro to the REST "architecture"
to web service tutorial guidelines api rest by hotho and 3 other people on 2006-04-04 16:11:47 copy

**Semantic Network Analysis of Ontologies**
Bettina Hoser and Andreas Hotho and Robert Jäschke and Christoph Schmitz and Gerd Stumme. *Proceedings of the 3rd European Semantic Web Conference \emph{(accepted for publication)}* (2006)
to web 2006 social ontology myown semantic analysis network sna by hotho and 1 other person on 2006-04-06 21:32:23 pick copy URL BibTeX

**Fig. 6.2** detail showing a single bookmark post

**Fig. 6.3** detail showing a single publication post

Since then, BibSonomy has rapidly grown and nowadays serves several thousand users – making it one of the top three social publication management systems.

## 6.4 Architecture

In this section we describe the architecture of BibSonomy's tag recommendation framework. BibSonomy itself is a web application based on the Java Servlet Technology[5] and a MySQL database.[6] An overview on its architecture and design can be found in [1] and [9].

### 6.4.1 Overview

Figure 6.4 gives an overview on the components of BibSonomy involved in a recommendation process. The web application receives the user's HTTP

---

[5] http://java.sun.com/products/servlets
[6] http://www.mysql.com/

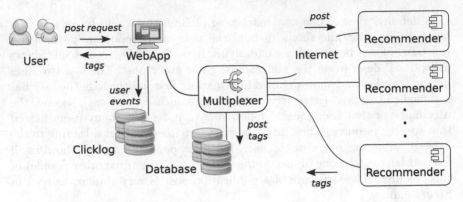

**Fig. 6.4** The involvement of BibSonomy's components in a schematic recommendation process.

request and queries the multiplexer (cf. Section 6.5.2) for a recommendation – which provides post information like URL, title, user name, etc. Besides, click events are logged in a database (see Section 6.4.3). The multiplexer then requests the active recommenders to produce recommendations and selects one of the results. The suggested tags and the post are then logged in a database and the selected recommendation is returned to the user.

### 6.4.2 Recommender Interface

One central element of the framework is the recommender interface. It specifies which data is passed from a recommendation request to one of the implemented recommenders and how they shall return their result. Figure 6.5 shows the UML class diagram of the *TagRecommender* interface one must implement to deliver recommendations to BibSonomy. We decided to keep

| <<interface>> |
| :---: |
| **TagRecommender** |
| + getRecommendedTags(post : Post<? extends Resource>) :      SortedSet<RecommendedTag> |
| + addRecommendedTags(      recommendedTags : Collection<RecommendedTag>,      post : Post<? extends Resource>) |
| + setFeedback(post : Post<? extends Resource>) |
| + getInfo() : String |

**Fig. 6.5** The UML class diagram of the tag recommender interface.

the interface as simple as possible by requiring only four methods, building on BibSonomy's existing data model (Post, Tag, etc.) and adding as few classes as possible (RecommendedTag, RecommendedTagComparator).

The *getRecommendedTags* method returns – given a post – a sorted set of tags; *addRecommendedTags* adds to a given (not necessarily empty) collection of tags further tags. Since – given a post and an empty collection – *addRecommendedTags* should return the same result as *getRecommended-Tags*, the latter can be implemented by delegation to the former. Nonetheless, we decided to require both methods to cover the simple 'give me some tags' case as well as more sophisticated usage scenarios (think of 'intelligent' collection implementations which could be handed to *addRecommendedTags*, or a recommender which improves given recommendations).

The post given to both methods contains data like URL, title, description, date, user name, etc. that will later be stored in the database and that the recommender can use to produce good recommendations. It might also contain tags, i. e., when the user edits an existing post or when he has already entered some tags and requests new recommendations. Implementations could use those tags to suggest different tags or to improve their recommendation.

With the *setFeedback* method, the final post (including the tags) as the user stored it in the database is given to the recommender such that it can measure and potentially improve its performance. Additionally, the *postID* introduced in Section 6.4.3 is contained in the post (as well as in the posts given to the first two methods) such that the recommender can connect the post with the recommended tags it provided.

Finally, the *getInfo* method allows the programmer to provide some information describing the recommender. This can be used to better identify recommenders or can be shown to the user.

Two further classes augment the interface: The *RecommendedTag* class basically extends the *Tag* class of the BibSonomy data model (cf. [9]) by adding floating point *score* and *confidence* attributes. A corresponding *RecommendedTagComparator* can be used to compare tags, e. g., for sorted sets. It first checks textual equality of tags (ignoring case) and then sorts them by score and confidence. Consequently, tags with equal names are regarded as equal.

Our implementation is based on Java. All described classes are contained in the module *bibsonomy-model*, which is available online as a Java archive in a Maven2 repository.[7] However, implementations are not restricted to Java – using the remote recommender (see Section 6.5.1.3) one can implement a recommender in any language which is then integrated using XML over HTTP requests.

---

[7] http://dev.bibsonomy.org/maven2/org/bibsonomy/bibsonomy-model/

### 6.4.3 Logging

For further evaluating the performance of the available tag recommenders, we store in a database for each recommendation process the corresponding bookmark or publication post as well as each recommender's recommendation, identified by a unique *recommendationID*. Furthermore, the applied selection strategy together with the selected recommenders and tags are stored.

Several recommendation requests may refer to a single posting process (i. e., when the user pressed the 'reload' button or forgot to enter a required field). For identifying these correspondences, a random identifier (*postID*) is generated whenever a post or editing process is started and retains valid until the corresponding post is finally stored in the database. This *postID* is mapped to each corresponding *recommendationID*. At storage time, the *postID* together with the corresponding user name, time stamp and the intra hash identifying the resource is stored. This connects each post of each user with all referring recommendations and vice versa.

Additionally, the user interaction is tracked by logging mouse click events using JavaScript. Each click on one of BibSonomy's web pages is logged using AJAX into a separate logging table. Information like the shown page, the DOM path of the clicked element, the underlying text, etc. is stored.[8]

## 6.5 Recommender Implementations

In this section we give an overview on some basic recommender implementations we realized within the framework. They serve as building blocks for more complex recommenders, e. g., the ones we present in Section 6.5.3.

### 6.5.1 Meta Recommender

*Meta* or *hybrid recommenders* [2] do not generate recommendations on their own but instead call other recommenders and modify or merge their results. Since they also implement the *TagRecommender* interface, they can be used like any other recommender. More formally, given $n$ recommendations $\hat{T}_{u,r}^1, \ldots, \hat{T}_{u,r}^n$ and corresponding scoring functions $\hat{s}^1, \ldots, \hat{s}^n$, a meta recommender produces a merged recommendation $\hat{T}_{u,r}$ with scoring function $\hat{s}$. The underlying design pattern known from software architecture is that of a *Composite* [5].

---

[8] Note that users can disable logging on the settings page, thus not all posting processes yield clicklog events.

As we will see in Section 6.5.3, meta recommenders allow the building of complex recommenders from simpler ones and thus simplify implementation and testing of algorithms and even stimulate development of new methods. Furthermore, they allow for flexible configuration, since their underlying recommenders can be exchanged at runtime. This section introduces the meta recommenders that are currently used in the framework.

### 6.5.1.1 First Weighted By Second

As an example of a cascade hybrid, the idea behind this recommender is to re-order the tags of one recommendation using scores from another recommendation. More precisely, given recommendations $\hat{T}_{u,r}^1$ and $\hat{T}_{u,r}^2$ and corresponding scoring functions $\hat{s}^1$ and $\hat{s}^2$, this recommender returns a recommendation $\hat{T}_{u,r}$ with scoring function $\hat{s}$, which contains all tags from $\hat{T}_{u,r}^1$ which appear in $\hat{T}_{u,r}^2$ (with $\hat{s}(t) := \hat{s}^2(t)$) plus all the remaining tags from $\hat{T}_{u,r}^1$ (with lower $\hat{s}$ but respecting the order induced by $\hat{s}^1$). If $\hat{T}_{u,r}^1$ does not contain enough recommendations, $\hat{T}_{u,r}$ is filled by the not yet used tags from $\hat{T}_{u,r}^2$ – again with $\hat{s}$ being lower than for the already contained tags and respecting the order induced by $\hat{s}^2$.

### 6.5.1.2 Weighted Merging

This weighted hybrid recommender enables merging of recommendations from different sources and weighting of their scores. Given $n$ recommendations $\hat{T}_{u,r}^1, \ldots, \hat{T}_{u,r}^n$, corresponding scoring functions $\hat{s}^1, \ldots, \hat{s}^n$, and (typically fixed) weights $\rho^1, \ldots, \rho^n$ (with $\sum_{i=1}^n \rho^i = 1$), the weighted merging recommender returns a recommendation $\hat{T}_{u,r} := \bigcup_{i=1}^n \hat{T}_{u,r}^i$ and a scoring function $\hat{s}(t) := \sum_{i=1}^n \rho^i \hat{s}^i(t)$ (with $\hat{s}^i(t) := 0$ for $t \notin \hat{T}_{u,r}^i$).

### 6.5.1.3 Remote Recommender

The remote recommender retrieves recommendations from an arbitrary external service using HTTP requests in REST-based [4] interaction. Therefore, it uses the XML schema of the BibSonomy REST API.[9] This recommender has three advantages: it allows us to distribute the recommendation work over several machines, it opens the framework to include recommenders from auxilliary partners, and it enables programming language independent interaction with the framework.

---

[9] http://www.bibsonomy.org/help/doc/xmlschema.html

To simplify implementation and integration of external recommenders, we provide an example web application needing almost no configuration to include a custom Java recommender.[10]

## 6.5.2 Multiplexing Tag Recommender

Our framework's technical core component is the so called *multiplexing tag recommender* (see Figure 6.4). Implementing BibSonomy's tag recommender interface, it provides the web application with tag recommendations by querying one of the configured recommenders. Furthermore, the multiplexer logs all recommendation requests and each recommender's corresponding result in a database (see Section 6.4.3). For this purpose, every tag recommender is registered during startup and assigned to a unique identifier.

Whenever the *getRecommendedTags* method of the multiplexer is invoked, the corresponding recommendation request is delegated to each available recommender, spawning a separate thread for each recommender. After a timeout period of 100 ms, one of the collected recommendations is selected, applying a preconfigured *selection strategy*. For our evaluation procedure we implemented a '*sampling without replacement*' strategy which randomly chooses exactly one recommender and returns all of its recommended tags. If the user requests recommendations more than once during the same posting process (e. g., by using the 'reload' button), the strategy selects recommendations from a recommender the user has not yet seen during this process.

## 6.5.3 Example Recommender Implementations

Using the proposed framework, we implemented several recommendation methods. Two of them were active in BibSonomy during the evaluation period in Section 7.2. Both build upon the meta recommenders described in Section 6.5.1 and simpler recommenders which we describe only briefly because they are fairly self-explanatory. The short names in parentheses are for later reference.

### 6.5.3.1 Most Popular $\rho$-Mix (MP$\rho$-mix)

Motivated by the good results of mixing tags which often have been attached to the resource with tags the user has often used (cf. Section 3.1.1), we implemented a variant of the *most popular $\rho$-mix* recommender described

---

[10] http://dev.bibsonomy.org/maven2/org/bibsonomy/bibsonomy-recommender-servlet

in [12]. Another factor was its efficient computability which can be supported by appropriate tables and indexes in the database. We set the parameter $\rho$ of this recommender to $\rho = 0.6$ for the evaluation in Section 7.2. The recommender has been implemented as a combination of three recommenders:

1. the *most popular tags by resource* (see Eq. 3.2) recommender which returns the $k$ tags $\hat{T}^1_{u,r}$ which have been attached to the resource most often (with $\hat{s}^1(t) := \frac{|Y \cap U \times \{t\} \times \{r\}|}{|Y \cap U \times T \times \{r\}|}$, i.e., the relative tag frequency),
2. the *most popular tags by user* recommender which returns the $k$ tags $\hat{T}^2_{u,r}$ the user has used most often (with $\hat{s}^2(t) := \frac{|Y \cap \{u\} \times \{t\} \times R|}{|Y \cap \{u\} \times T \times R|}$, i.e., the relative tag frequency), and
3. the *weighted merging* meta recommender described in Section 6.5.1.2 which merges the tags of the two former recommenders, with weights $\rho^1 = \rho = 0.6$ and $\rho^2 = 1 - \rho = 0.4$.

### 6.5.3.2 Title Tags Weighted by User Tags (TbyU)

Inspired by the first recommender implemented in BibSonomy [8] and by similar ideas in [14], we implemented a recommender which scores tags extracted from the resource's title using the frequency of the tags used by the user. Technically, this is again a combination of three recommenders:

1. a simple *content based recommender*, which extracts $k$ tags $\hat{T}^1_{u,r}$ from the title of a resource, cleans them and checks against a multilingual stopword list,
2. the *most popular tags by user* recommender as described in the previous section – here returning *all* tags $\hat{T}^2_{u,r}$ the user has used (by setting $k = \infty$), and
3. the *first weighted by second* meta recommender described in Section 6.5.1.1 which weights the tags from the content based recommender by the frequency of their usage by the user as given by the second recommender.

### 6.5.3.3 Other

Besides the simple recommenders introduced along the MP$\rho$-mix and TbyU recommender, we have implemented recommenders for testing purposes (a *fixed tags recommender* and a *random tags recommender*), a recommender which proposes tags from a web page's HTML meta information keywords, as well as a recommender using the FolkRank algorithm.

More complex recommenders can be thought of, e.g., a nested *first weighted by second* recommender, whose first recommender is a *weighted merging* meta recommender merging the suggestions from a *content based recommender* and a *most popular tags by resource* recommender and then scoring the tags by the scores from the *most popular tags by user* recommender.

## 6.6 Further Reading

Although having a different recommendation target (resources rather than tags), the REFEREE framework described by Cosley et al. [3] is most closely related to our work. It provided recommendations for the CiteSeer (formerly ResearchIndex) digital library. REFEREE recommends scientific articles to users of ResearchIndex while they search and browse. An open architecture allows researchers to integrate their methods into REFEREE. Besides the different recommendation target, the focus of the paper is more on the evaluation of several different strategies than on the details of the framework.

A powerful, open, and well documented framework for recommendations is MyMediaLite [6]. It addresses the rating prediction and item prediction scenario in collaborative filtering and has a focus on explicit user ratings and non re-occurring items, e. g., like in a movie recommendation scenario where one does not recommend movies the user has already seen. This is in contrast to tag recommendations, where re-occurring tags are a crucial requirement of the system.

Another recommendation framework is TasteKeeper [7] from Sun Microsystems' AURA project.[11] Despite not having been described in the literature, it has a strong focus on collaborative filtering algorithms. Finally, the machine learning library Apache Mahout [15] also contains implementations of collaborative filtering and user/item based recommenders that can be run in parallel in a distributed setup.

## References

1. Dominik Benz, Andreas Hotho, Robert Jäschke, Beate Krause, Folke Mitzlaff, Christoph Schmitz, and Gerd Stumme. The social bookmark and publication management system BibSonomy. *The VLDB Journal*, 19(6):849–875, December 2010.
2. Robin Burke. Hybrid recommender systems: Survey and experiments. *User Modeling and User-Adapted Interaction*, 12(4):331–370, November 2002.
3. Dan Cosley, Steve Lawrence, and David M. Pennock. REFEREE: an open framework for practical testing of recommender systems using Re-

---

[11] Advanced Universal Recommendation Architecture Project http://kenai.com/projects/aura/

searchIndex. In *VLDB '02: Proceedings of the 28th international confer-ence on Very Large Data Bases*, pages 35–46. VLDB Endowment, 2002.

4. Roy T. Fielding. *Architectural Styles and the Design of Network-based Software Architectures*. PhD thesis, University of California, Irvine, 2000.

5. Erich Gamma, Richard Helm, and Ralph E. Johnson. *Design Patterns. Elements of Reusable Object-Oriented Software*. Addison-Wesley Long-man, Amsterdam, 1st edition, 1995.

6. Zeno Gantner, Steffen Rendle, Christoph Freudenthaler, and Lars Schmidt-Thieme. MyMediaLite: A free recommender system library. In *Proceedings of the 5th ACM Conference on Recommender Systems (Rec-Sys 2011)*, New York, NY, USA, 2011. ACM.

7. Steve Green and Jeff Alexander. The advanced universal recommenda-tion architecture (AURA) project. http://www.tastekeeper.com/, 2008.

8. Jens Illig. Entwurf und Integration eines Item-Based Collaborative Fil-tering Tag Recommender Systems in das BibSonomy-Projekt. Project report, Fachgebiet Wissensverarbeitung, Universität Kassel, 2006.

9. Robert Jäschke. *Formal Concept Analysis and Tag Recommendations in Collaborative Tagging Systems*, volume 332 of *Dissertationen zur Künstlichen Intelligenz*. Akademische Verlagsgesellschaft AKA, Heidel-berg, Germany, January 2011.

10. Robert Jäschke, Folke Eisterlehner, Andreas Hotho, and Gerd Stumme. Testing and evaluating tag recommenders in a live system. In *RecSys '09: Proceedings of the 2009 ACM Conference on Recommender Systems*, pages 369–372, New York, NY, USA, October 2009. ACM.

11. Robert Jäschke, Folke Eisterlehner, Andreas Hotho, and Gerd Stumme. Testing and evaluating tag recommenders in a live system. In Dominik Benz and Frederik Janssen, editors, *Workshop on Knowledge Discovery, Data Mining, and Machine Learning*, pages 44–51, September 2009.

12. Robert Jäschke, Leandro Marinho, Andreas Hotho, Lars Schmidt-Thieme, and Gerd Stumme. Tag recommendations in social bookmarking systems. *AI Communications*, 21(4):231–247, 2008.

13. Leslie Lamport. *LATEX: A Document Preparation System*. Addison-Wesley, 1986.

14. M. Lipczak. Tag recommendation for folksonomies oriented towards individual users. In Andreas Hotho, Beate Krause, Dominik Benz, and Robert Jäschke, editors, *ECML PKDD Discovery Challenge 2008 (RSDC'08)*, pages 84–95, 2008.

15. Sean Owen, Robin Anil, Ted Dunning, and Ellen Friedman. *Mahout in Action*. Manning Publications, 1st edition, 2011.

16. Oren Patashnik. BibTeXing, 1988. (Included in the BIBTEX distribution).

# Chapter 7
# Online Evaluation

The multiplexing tag recommender of BibSonomy allows for comparisons of different tag recommenders in a realistic real-life setting. We show in this chapter, which kind of evaluation the framework allows and how recommenders perform in practice. We begin with an introduction of the evaluation setting (Section 7.1) and then present in Section 7.2 a case study involving two simple recommendation methods. Finally, in Section 7.3, the online recommendation task of the ECML PKDD Discovery Challenge 2009 is presented which was performed and evaluated using the framework.

## 7.1 Evaluation Setting

### 7.1.1 Metrics and Protocols

We used precision, recall, and the F1 measure for evaluation (cf. Section 5.1.1) and determined the winner of the Discovery Challenge by the best F1 measure when regarding the *first five tags* of the recommendation $\hat{T}_{u,r}$. The change of precision and recall for an increasing number of recommended items can be seen in plots like the one shown in Figure 7.1. It shows the typical behaviour of recommender systems: the more items are recommended, the better the recall but the worse the precision becomes.

### 7.1.2 Preprocessing and Cleansing

Before comparing the recommended tags $\hat{T}_{u,r}$ with the tags $T_{u,r}$, the user $r$ chose to tag the resource $r$ with, we clean the tags in both sets according to the Java method *cleanTag* shown in Algorithm 7.1. This means, we ignore the case of tags and remove all characters which are neither numbers nor letters.[1] Since we assume all characters to be UTF-8 encoded, the method will *not* remove umlauts and other non-latin characters. We also employ Unicode normalization to normal form KC[2] using *java.text.Normalizer*. Finally, we ignore tags which are 'empty' after normalization (i. e., they neither contained a letter nor number) or which are equal to the strings *imported*, *public*, *systemimported*, *nn*, *systemunfiled*. Thus, in the following we always regard cleaned tags.

---

**Algorithm 7.1** The Java method used to clean tags.

---

```
1  public String cleanTag(String tag) {
2      return Normalizer.normalize(tag, Normalizer.Form.NFKC).
3          replaceAll("[^0-9\\p{L}]+", "").
4          toLowerCase();
5  }
```

---

## 7.2 Case Study

In this section we show by means of the two simple recommenders introduced in Section 6.5.3 which kind of evaluation the tag recommendation framework of BibSonomy supports and how those two recommenders perform in practice. The analysis is based on data from posting processes between May 15th and June 26th 2009. Only public posts from users not flagged as spammer were taken into account.[3] Since tag recommendations are provided in the web application only when *one* resource is posted, posts originating from imports (e. g., Firefox bookmarks, or BIBTEX files) or BibSonomy's API are not contained in the analysis.

### 7.2.1 General Results

We start with some general numbers: In the analyzed period, 5,840 posting processes (3,474 for publications, 2,366 for bookmarks) have been provided with tag recommendations. The MP$\rho$-mix recommender served recommen-

---

[1] See also the documentation of *java.util.regex.Pattern* at http://download.oracle.com/javase/6/docs/api/java/util/regex/Pattern.html.

[2] http://www.unicode.org/unicode/reports/tr15/tr15-23.html

[3] Users can be flagged as spammers manually or by BibSonomy's spam detection framework [4].

dations for 2,935 postings, the TbyU recommender for 3,006. Their precision and recall is depicted in Figure 7.1. On the plotted curve, from left to right the number of evaluated tags increases from one to five. I. e., we first regard only the tag $t$ with the highest value $\hat{s}(t)$, then the two tags with highest $\hat{s}$, and so on. Thus, the more recommended tags are regarded, recall increases while precision decreases. In general, both precision and recall are rather low with the MP$\rho$-mix recommender performing better than the TbyU recommender.

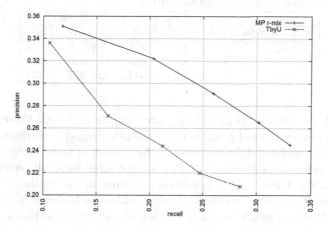

**Fig. 7.1** Recall and precision of the two deployed recommenders. The number of recommended tags increases from one on the left to five on the right as described in Section 5.1.1.

## 7.2.2 Influence of the 'reload' Button

Since users can request to reload recommendations when posting a resource, we here investigate the influence of the 'reload' button. Is the first recommendation sufficient or do users request another recommendation? Are recommendations which got replaced by the user pressing the 'reload' button worse than those shown last? Has one recommender more often been reloaded than the other?

In 767 (274 bookmark, 493 BibTeX) of the 5,840 posting processes the users requested to reload the recommendation. Thus, in around 13 % of all posting processes users requested another recommendation.

Recommendations from several recommenders can be displayed during one posting process. There is the recommendation which appears directly after loading the posting page (*first*), there are recommendations which appear after the user has pressed the 'reload' button, and there is the recommendation shown before the user finally saves the post (*last*). Thus, given a recommender

$\mathfrak{r}$, we can define the set $F_{\mathfrak{r}}$ to contain those posts, where the recommender $\mathfrak{r}$ showed the first tags, and $L_{\mathfrak{r}}$ as the set of posts where the recommender $\mathfrak{r}$ showed the last tags (i. e., before the post is finally stored). For each recommender $\mathfrak{r}$, we can then look at the sets $F_{\mathfrak{r}} \setminus L_{\mathfrak{r}}$, $L_{\mathfrak{r}} \setminus F_{\mathfrak{r}}$, and $F_{\mathfrak{r}} \cap L_{\mathfrak{r}}$. Posts where the user did not press the reload button are contained in both $F_{\mathfrak{r}}$ and $L_{\mathfrak{r}}$ and thus in $F_{\mathfrak{r}} \cap L_{\mathfrak{r}}$. Table 7.1 shows the result of our analysis.

**Table 7.1**  The influence of the 'reload' button.

| measure | #posts | | f1m@5 | |
|---|---|---|---|---|
| recommender $\mathfrak{r}$ | MP$\rho$-mix | TbyU | MP$\rho$-mix | TbyU |
| $F_{\mathfrak{r}} \setminus L_{\mathfrak{r}}$ | 337 | 319 | 0.258 | 0.270 |
| $L_{\mathfrak{r}} \setminus F_{\mathfrak{r}}$ | 331 | 363 | 0.380 | 0.364 |
| $F_{\mathfrak{r}} \cap L_{\mathfrak{r}}$ | 2,271 | 2,339 | 0.277 | 0.224 |

For both of the two deployed recommenders and for all three sets, the table shows the number of posts in the corresponding set, and the average f1-measure at the fifth tag.[4] As one can see, the number of posts where the reload button has not been pressed ($F_{\mathfrak{r}} \cap L_{\mathfrak{r}}$) is quite large for both recommenders (around $2,300$). There is also only little difference in the number of posts for the recommenders over the different sets, besides the higher number of posts for the TbyU recommender in the $L_{\mathfrak{r}} \setminus F_{\mathfrak{r}}$ set. It contains those posts, where the user requested to reload the recommendation and where the recommender at hand delivered the last recommendation. Thus, the TbyU recommender more often provided the last recommendation than the MP$\rho$-mix recommender.

The most noticeable observation is the good performance of both recommenders for the posts where a reload occurred and the recommender showed the last recommendations ($L_{\mathfrak{r}} \setminus F_{\mathfrak{r}}$). There, both precision and recall are much higher than for the other two sets. This suggests that the first suggestion was rather bad and caused the user to request another recommendation which indeed better fitted his needs. The worse values for the $F_{\mathfrak{r}} \setminus L_{\mathfrak{r}}$ set also support this thesis. A noteworthy difference between the two recommenders is the performance of the TbyU recommender for the $F_{\mathfrak{r}} \setminus L_{\mathfrak{r}}$ set which is better than its overall performance (i. e., on the $F_{\mathfrak{r}} \cap L_{\mathfrak{r}}$ set). This could be an indicator that those users which actively used the recommender (by pressing the 'reload' button) took better notice of this recommender's tag suggestions.

The usage of the 'reload' button seems to be a good indicator for the interest of the user in the recommendations. However, the data we gathered during the one month evaluation period is still rather sparse, thus no final conclusions can be drawn.

---

[4] We omit precision and recall, since whenever the f1m for one set was better/worse than for another set, precision and recall were better/worse, too.

### 7.2.3 Logged 'click' Events

Next we evaluate the data from the log which records when a user clicked on a recommended tag (cf. Section 6.4.3). Clicks are rather sparse: in only 1,061 (485 bookmarks, 576 publications) of the 5,840 posting processes, users clicked on a recommendation.

First, we want to answer the questions "How is clicking distributed over users?" and "Are there users which always/never click?". Figure 7.2 shows the

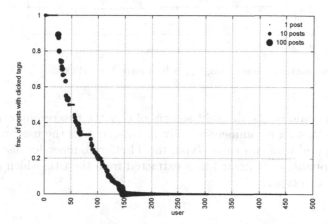

**Fig. 7.2** Users sorted by their fraction of click/noclick-posts. The y-axis depicts the fraction of posts where recomended tags were clicked. Each circle represents a user. As shown in the scale at the upper right corner, the size of each circle depicts the logarithm of the user's number of posts regarded for the analysis.

users sorted by the fraction of posting processes at which they have clicked on a recommended tag. The size of each circle depicts the logarithm of the user's number of posting processes incorporated into the analysis. Closer to the left are users which in almost all posting processes clicked on a recommendation; users closer to the right never clicked a recommended tag during posting. Only around 150 users clicked on a tag and half of the remaining users are represented by only one post. This could mean that only after some time users discover and use the recommendations. However, there are also some active users which almost never clicked on a recommendation.

In Figure 7.3 we see for each number of recommended tags (from one to five), the fraction of matches which stem from a click on the tag (instead of manual typing). For the TbyU recommender, around 35 % of the matches come from the user clicking on a tag. Thus, although users infrequently click on tags, a large fraction of the correctly recommended tags of that recommender has been clicked instead of typed. Why there is a difference of around 15 % between the two recommenders with a higher click fraction for the TbyU recommender (in contrast to its worse precision and recall) is not clear. One

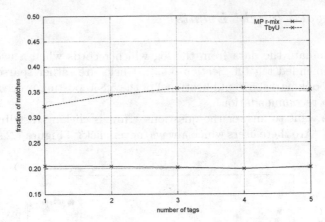

**Fig. 7.3** The fraction of matching tags which have been clicked.

explanation could be the different sources of tags the two recommenders use: while the MP$\rho$-mix recommender delivers popular tags the user might have used before and thus can easily type, the TbyU recommender also suggests new and probably complicated tags extracted from the title which are easier to click than to type.

### 7.2.4 Average F1-Measure per User

Which properties of a posting process could help a multiplexer strategy to smartly choose a certain recommender instead of randomly selecting one? We here focus on the user only – other characteristics could be likewise interesting (e.g., resource type or the recommended tags). Figure 7.4 shows the average f1m of the MP$\rho$-mix recommender versus the average f1m of the TbyU recommender for each of the 380 users[5] in the data. In the plot, each user is represented by a circle whose size depicts the logarithm of the user's number of posts.

The most interesting users are reflected by the circles farthest from the diagonal, i.e., those users who have a high f1m for one but a low f1m for the other recommender. As one can see, such users exist even at higher post counts. Once such a user is identified, one could primarily select recommendations from the user's preferred recommender, e.g., by increasing the probability for randomly selecting the recommender.

---

[5] Only users which got recommendations from *both* recommenders were taken into account.

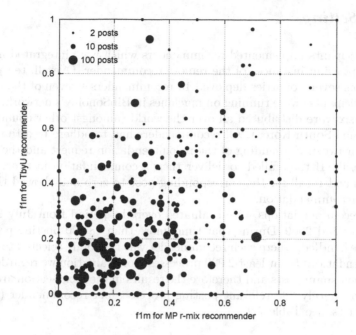

**Fig. 7.4** Average f1-measure for each user and recommender. Each circle represents a user as described for Figure 7.2.

## 7.3 The ECML PKDD Discovery Challenge 2009

Continuing the analysis presented in the previous section, we now focus on a larger setting: The framework was the cornerstone of the ECML PKDD[6] Discovery Challenge 2009 [3][7] where one task required the participants to deliver online recommendations to BibSonomy. This was a larger stress test for external recommenders and the framework itself. In this section we give a brief overview on the setting, the methods some recommenders used and the resulting recommendation performances.

---

[6] The *European Conference on Machine Learning and Principles and Practice of Knowledge Discovery and Data Mining* is according to its website (http://www. ecmlpkdd.org/content/past-conferences) the "largest European conference in these areas".

[7] http://www.kde.cs.uni-kassel.de/ws/dc09/

## 7.3.1 Setting

The participants implemented recommenders which were integrated into the framework using instances of the remote recommender. Overall, ten participants from seven countries deployed 13 recommenders – seven of them (from four participants) were running on machines in BibSonomy's network, the remaining six were distributed all over the world (amongst other countries, in Canada and South Korea). All recommenders had to adhere to a timeout of 1000 ms between the sending of the recommendation request and the arrival of the result. If they failed to deliver their recommendation in time, we set precision and recall for the corresponding post to zero and showed the user another recommendation.

The recommendations were evaluated over the period from July 29th to September 2nd 2009. During that time, more than 28,000 posting processes had to be handled, where each recommender was randomly selected to deliver recommendations for at least 2,000 processes. For evaluation we regarded only public, non-spam posts and therefore the results in the next section are based on approximately 380 relevant posting processes per recommender (for the exact counts, see Table 7.2).

**Table 7.2** The number of posts regarded for evaluation.

| recommender-id | 3 | 5 | 6 | 7 | 12 | 13 | 14 | 16 |
|---|---|---|---|---|---|---|---|---|
| #posts | 347 | 391 | 361 | 415 | 385 | 380 | 370 | 398 |

Although 13 recommenders participated in the online task, only eight of them managed to deliver results in at least 50 % of all requested posting processes. The remaining five recommenders answered only in less than 5 % of all cases and are thus ignored in Table 7.2 and in the figures and discussion following in Section 7.3.3.

## 7.3.2 Methods

Details on the tag recommendation methods evaluated during the challenge can be found in the proceedings [3]. Here we only briefly introduce the three best recommenders of the online task.

The winning recommender 6 [5] uses a method based on the combination of tags from the resource's title, tags assigned to the resource by other users and tags in the user's profile. The system is composed of six recommenders and the basic idea is to augment the tags from the title by related tags extracted from two tag-tag–co-occurrence graphs and from the user's profile and then rescore and merge them.

Recommender 3 [6] performs so called "Feature Driven Tagging" by extracting and weighting features like words, ids, hashes, phrases from the resources. Each feature then generates a list of tags. The weight of the features is estimated using TF×IDF and TF×ITF (term frequency × inverted document frequency and term frequency × inverted tag frequency – see [1]); the tags of the features are determined using co-occurrence counts, mutual information, and $\chi^2$ statistics.

For recommender 5, Cao et al. [2] divide the posts in four categories, depending on the case if the user or resource of the post is known or not. Then, for each category they learn a model to rank the tags using a ranking SVM. To augment the available tags for posts (besides the full text and the tags of the resource), the authors use post-content similarity and $k$-Nearest-Neighbors.

## 7.3.3 Results

**Overall Performance.** First, we have a look at precision and recall of each recommender in the evaluation mode relevant for the challenge (Figure 7.5(a)). For a posting process in which the recommender could not deliver a recommendation in time, precision and recall were set to zero. In this setting, recommender 6 [5] is the clear winner with an f1m of 0.205 for five tags. The performance of the remaining methods varies between an f1m of 0.030 and 0.171 for five tags – all those recommenders have a recall of less than 0.2.

**Influence of the Recommendation Latency.** If we disregard the timeout limit of 1000 ms and also evaluate the suggestions which came later (cf. Figure 7.5(b)), we get a different picture. Of course, all recommenders improve – but in particular recommender 14 gains both precision and recall. This can be explained by the latency plot shown in Figure 7.6. It shows for each recommender the latency of the delivered recommendations for the selected posting processes, ordered in ascending order by latency. The curves do not reach 100 % because in some cases the recommenders did not deliver a result at all. One can see that recommender 14 returned a suggestion in almost as many posts as the winning recommender 6. However, only 20 % of the posts were delivered in time – in contrast to almost 80 % of the posts for recommender 6. Consequently, timeouts are a serious issue in this setting – with a timeout of 2000 ms, the competition for the best performance would have been much closer. Nevertheless, a timeout of 2000 ms would be too long for recommendations which shall be shown after loading the page. One should also note that in principle network latency was not an issue since the winning recommender was located in Canada.

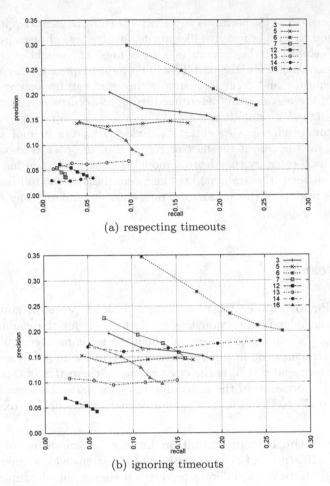

(a) respecting timeouts

(b) ignoring timeouts

**Fig. 7.5** Recall and precision of the deployed recommenders.

**Comparison with Offline Results.** Before the participants tested their recommenders in the online setting, most of them performed an offline evaluation against a dataset from BibSonomy. Interestingly, some recommenders gained better results in the online challenge than in the offline challenge (see Figure 7.7). Without going too much into detail, one explanation could be the fact that in the online challenge the user actually *saw* the recommender's suggestion and thus had the chance to utilize it. This suggests that users actively used the recommendations and are indeed influenced by them.

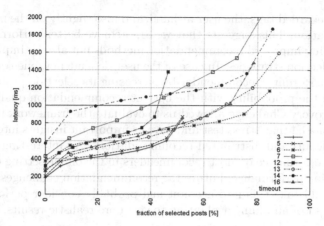

**Fig. 7.6** Latency of the recommenders. No curve reaches 100 % because none of the recommenders delivered results for all posting processes they have been selected for.

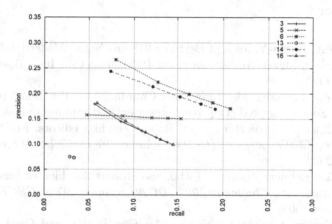

**Fig. 7.7** Performance in the offline task (recommenders 7 and 12 did not participate).

## 7.4 Conclusion

In this chapter we evaluated the tag recommendation framework that we developed for BibSonomy. It allows us to not only integrate and judge recommendations from various sources but also to develop clever selection strategies. A strength of the framework is its ability to log all steps of the recommendation process and thereby making it traceable. E. g., the diagrams and tables presented in this chapter are automatically generated and will be integrated in a web application for analysing and controlling the framework and its recommenders.

As the results in Section 7.2 show, there is no clear picture which of the two recommendation methods performs better. There is a dependency on the number of regarded tags, the user at hand, and also slightly on the moment of recommendation. This suggests that we can achieve better performance not only by adding improved recommendation methods but also by implementing adaptive selection strategies. In case of the user dependency, one could prefer the better performing recommender by increasing its selection probability or even couple the probability with the current recommendation quality.

The Discovery Challenge allowed us to evaluate the framework in a larger setting. It passed that stress test and gave us important insights into the handling of timeouts and distributed recommendations. An interesting finding is the better performance of most recommenders in the online setting compared to their offline performance. Future tag recommendation challenges and evaluations should take this into account and probably consider performing an online instead of an offline evaluation to get more realistic results.

# References

1. Ricardo A. Baeza-Yates and Berthier Ribeiro-Neto. *Modern Information Retrieval.* Addison-Wesley Longman Publishing Co., Inc., Boston, MA, USA, 1999.
2. Hao Cao, Maoqiang Xie, Lian Xue, Chunhua Liu, Fei Teng, and Yalou Huang. Social tag prediction base on supervised ranking model. In Folke Eisterlehner, Andreas Hotho, and Robert Jäschke, editors, *ECML PKDD Discovery Challenge 2009 (DC09)*, volume 497 of *CEUR-WS.org*, pages 35–48, 2009.
3. Folke Eisterlehner, Andreas Hotho, and Robert Jäschke, editors. *ECML PKDD Discovery Challenge 2009 (DC09)*, volume 497 of *CEUR-WS.org*, September 2009.
4. Beate Krause, Christoph Schmitz, Andreas Hotho, and Gerd Stumme. The anti-social tagger – detecting spam in social bookmarking systems. In *AIRWeb '08: Proceedings of the 4th International Workshop on Adversarial Information Retrieval on the Web*, pages 61–68, New York, NY, USA, April 2008. ACM.
5. Marek Lipczak, Yeming Hu, Yael Kollet, and Evangelos Milios. Tag sources for recommendation in collaborative tagging systems. In Folke Eisterlehner, Andreas Hotho, and Robert Jäschke, editors, *ECML PKDD Discovery Challenge 2009 (DC09)*, volume 497 of *CEUR-WS.org*, pages 157–172, 2009.
6. Xiance Si, Zhiyuan Liu, Peng Li, Qixia Jiang, and Maosong Sun. Content-based and graph-based tag suggestion. In Folke Eisterlehner, Andreas Hotho, and Robert Jäschke, editors, *ECML PKDD Discovery Challenge 2009 (DC09)*, volume 497 of *CEUR-WS.org*, pages 243–260, 2009.

# Chapter 8
# Conclusions

In this chapter we close the book with a summary, a discussion, and future research directions.

## 8.1 Summary

The advent of social tagging systems changed the way people create and consume content in the Web. Those systems represent a human computation paradigm with enormous potential to address problems that computer programs cannot yet tackle on their own, since the tagging of resources is done by human beings, who understand the content of the resources. In this way, tags can serve as rich indexes for finding and organizing content, not only by humans, but also by computer programs. Due to the increasing popularity of these systems, information overload rapidly becomes a problem. Recommender systems proved to be well suited for this kind of problem in the past and are thus a potential solution for tackling the information overload in the Web's next generation. Summaries of the topics presented in this book are as follows:

- The data structure of folksonomies, stressing the differences in comparison to the ones used by traditional recommender systems.
- A formal description of recommender systems as well as their different tasks and set-ups.
- Recommendation algorithms that:
  - count frequencies of (co-)occurrences of given entities in the data,
  - operate on two dimensional projection matrices,
  - operate directly on the ternary relational data of folksonomies,
  - exploit external sources of information.
- Evaluation metrics and protocols.

- A case study presenting the implementation and deployment of recommender systems in BibSonomy, stressing the challenges and requirements for doing so.

## 8.2 Discussion and Outlook

Many STS, such as BibSonomy,[1] provide snapshots of their datasets for research purposes, which together with the ECML PKDD Discovery Challenge 2009 [1] has contributed to turn the problem of tag recommendation into an active area of research.

The ternary relational data of folksonomies open new and interesting research directions in recommender systems since most of the existing methods are specialized on binary relational data. Even though we have presented several different ways of handling ternary relations when designing recommenders for STS, new progress in this field will certainly continue to blossom. A particularly appealing research direction concerns investigating multi-mode recommendation frameworks, such as the FM presented in Section 4.1.2.4, that support easy switching between different recommendation tasks and number of modes.

Although we have focused on the plain recommendation problem, without attributes or other external sources of information (sometimes called contextual information), there is an increasing interest on context-aware models able to capture and exploit different contexts at the same time, such as content (e. g., text, image, and video), friendship relations, spatio-temporal and semantic metadata. Once more, FM like models, appear as a very promising research direction in this area.

Another important research direction is recommender systems for mobile environments. Many Web 2.0 applications are now deployed on mobile devices, such as smart phones and tablets, where the computational resources are limited. Thus, it is essential to develop recommendation algorithms that can operate efficiently in such limited environments.

Other topics that were not covered in this book, but are nevertheless interesting research directions, concern, for example, recommendations' novelty and serendipity, e. g., tags that are potentially interesting but not obvious; cold-start scenarios, such as the new user/resource problem; and cross domain recommendations. In all, we hope the work here presented can inspire further explorations in these research areas.

---

[1] http://www.kde.cs.uni-kassel.de/bibsonomy/dumps/

# References

1. Folke Eisterlehner, Andreas Hotho, and Robert Jäschke, editors. *ECML PKDD Discovery Challenge 2009 (DC09)*, volume 497 of *CEUR-WS.org*, September 2009.